AF594944

Chromosome Manipulation for Plant Breeding Purposes

Chromosome Manipulation for Plant Breeding Purposes

Editor

Pilar Prieto

MDPI • Basel • Beijing • Wuhan • Barcelona • Belgrade • Manchester • Tokyo • Cluj • Tianjin

Editor
Pilar Prieto
Institute for Sustainable Agriculture
Spain

Editorial Office
MDPI
St. Alban-Anlage 66
4052 Basel, Switzerland

This is a reprint of articles from the Special Issue published online in the open access journal *Agronomy* (ISSN 2073-4395) (available at: https://www.mdpi.com/journal/agronomy/special_issues/chromosome_plant_breeding).

For citation purposes, cite each article independently as indicated on the article page online and as indicated below:

LastName, A.A.; LastName, B.B.; LastName, C.C. Article Title. *Journal Name* **Year**, *Volume Number*, Page Range.

ISBN 978-3-0365-0024-9 (Hbk)
ISBN 978-3-0365-0025-6 (PDF)

Contents

About the Editor

Pilar Prieto (Ph.D. in Biology) is a Research Scientist at the Institute for Sustainable Agriculture and a member of the Spanish National Research Council (CSIC). She graduated from the University of Córdoba in 1998 where she was later awarded her Ph.D. in Biological Sciences in 2002. Her main research focus is on meiosis for plant breeding purposes using cytogenetics and confocal microscopy tools. Specific topics are the use of biotechnological tools for plant meiosis studies to promote interspecific recombination in cereals, chromosome manipulation in cereals to facilitate introgression of desirable agronomic traits in crops (mainly in wheat) from related wild species, and the study of genome organization and nuclear architecture in plants, among others. She is Professor at the University of Córdoba, where she delivers lectures for the Master in Biotechnology, and has also served on the CSIC Life Sciences Committee since November 2018.

Editorial

Chromosome Manipulation for Plant Breeding Purposes

Pilar Prieto

Plant Breeding Department, Institute for Sustainable Agriculture, Agencia Estatal Consejo Superior de Investigaciones Científicas (CSIC), Alameda del Obispo s/n, Apartado 4084, 14080 Córdoba, Spain; pilar.prieto@ias.csic.es; Tel.: +34-957-499-293

Received: 27 October 2020; Accepted: 29 October 2020; Published: 2 November 2020

Abstract: The transfer of genetic variability from related species into crops has been a main objective for decades in breeding programs. Breeders have used interspecific genetic crosses and alien introgression lines to achieve this goal, but the success is always dependent on the interspecific chromosome associations between the alien chromosomes and those from the crop during early meiosis. In this Special Issue, the strength of chromosome manipulation in a breeding framework is revealed through research and review papers that combine molecular markers, cytogenetics tools and other traditional breeding techniques. The papers and reviews included in this Special Issue "Chromosome manipulation for plant breeding purposes" describe the development and/or characterization of new plant material carrying desirable traits and the study of chromosome associations and recombination during meiosis. New tools to facilitate the transfer of desired traits from a donor species into a crop can be developed by expanding the knowledge of chromosome associations during meiosis.

Keywords: meiosis; chromosome engineering; chromosome pairing; non-homologous recombination; cytogenetics; alien chromosome; polyploidy; aneuploidy

---

1. Introduction

The capacity to exploit the potential of wild relatives carrying beneficial traits is a key goal in breeding programs. However, it depends on the opportunity of the chromosomes from the crop and the wild species in interspecific crosses to recognize, associate and undergo crossover formation during meiosis, the cellular process responsible for producing gametes carrying half the genetic content of the parent cells. Unfortunately, in most of the cases, a barrier prevents successful hybridization between the wild and the crop chromosomes. Some other approaches, such as radiation and gametocidal genes, have been used to transfer resistance genes or chromosomal segments with desirable genes from related species into a crop such as wheat. Unluckily, both approaches generate arbitrary chromosome breaks and fusion and most of the translocations occur between non-homologous chromosomes, generating deficiencies or genetic duplications that are not compensated, and, therefore, do not have interest as tools for breeding. It seems interesting to develop chromosome manipulation methods affecting homologous recombination to produce genetic introgressions that can be genetically equilibrated and transmitted to the descendance. Understanding the mechanisms controlling chromosome associations during meiosis are, therefore, of great interest in plant breeding. Examining how chromosomes recognize, associate in pairs, synapse and recombine, which are prerequisites for the balanced segregation of half-bivalents during meiosis, will allow chromosome manipulation to introduce genetic variability from related species into a crop.

Several genetic stocks such as interspecific hybrids, natural and synthetic polyploids and introgression lines derived from allopolyploids, among others, can be powerful tools to transfer genetic traits from one species into another and for meiosis studies. For example, an extra pair of alien chromosomes in the full genome of a crop species has been often used as a first step in breeding programs to introduce genetic variation from the secondary gene pool. Moreover, such introgression lines are also essential in the study of interspecific genetic interactions, in the chromosomal location of genetic markers and in the study of chromosome structure and behaviour in somatic and meiotic cells.

Through several research papers and reviews, this Special Issue of Agronomy describes the use of chromosome introgressions as an excellent tool to transfer several characters mainly into cereal crops such as wheat, rye and triticale and their use in meiosis studies with the aim of promoting interspecific chromosome rearrangements and recombination. Other genetic stocks are also studied in this Special Issue for plant breeding purposes.

2. Chromosome Manipulation within the Triticeae Tribe

The *Triticeae* tribe is included in the *Poaceae* (*Gramineae*) family and contains grain crops of world-wide economic importance such as wheat (*Triticum*), barley (*Hordeum*) and rye (*Secale*), as well as a profuse number of grasses used for animal feed or rangeland protection. Bread wheat is one of the most important crops in the world. Breeders have been using related species as genetic donors with the aim of widening the genetic basis of a crop and getting, for example, wheat cultivars better adapted to specific agro-climatic conditions or carrying resistance to pests. *Hordeum chilense*, a perennial diploid wild barley that has an extremely high potential for wheat breeding, has been used to develop wheat-*H. chilense* chromosome $2H^{ch}$ introgression lines to be used for improving grain quality [1]. Chromosome $2H^{ch}$ carries endosperm carotenoid-related genes, which can contribute to improve seed carotenoid content in wheat. These authors have used a wheat disomic addition line carrying $2C^{c}$ chromosome from *Aegilops cylindrica* Host. to induce chromosome reorganizations between $2H^{ch}$ and wheat chromosomes. New wheat-*H. chilense* recombinants that can be useful for studying the effect of chromosome $2H^{ch}$ in grain quality have been identified using *in situ* hybridisation and molecular markers.

Hordeum chilense has been also used to evaluate the variability for the *Wx* gene that encodes a waxy protein (a granule-bound starch synthase I), responsible for amylose synthesis [2]. Authors have located this *Wx* gene on the long arm of $7H^{ch}$ chromosome and have identified two different alleles of this gene in *H. chilense*. This genetic variability could be transferred into wheat by introgressing $7H^{ch}L$ chromosome segments, as the current variability present in wheat cultivars is not very wide.

Several genes contribute to the bread-making quality in wheat such as glutenins, the wheat bread-making (*wbm*) gene and the presence of the 1BL.1RS translocation, among others. Little is known about the origin of the *wbm* gene and its genomic organization [3]. These authors examined *wbm* evolution and demonstrated the location of this gene in 7AL chromosome. Thus, it was possible to identify triticale lines carrying *wbm* to be used in the improvement of the bread-making quality of this crop.

Chromosome introgressions are also used to transfer resistance against pathogens. In triticale (×*Triticosecale* Wittmack) new pathogens have evolved, moving from wheat and rye into triticale due to the increment of the harvested area. A screening of robertsonian translocations in the progeny of triticale lines carrying monosomic substitutions of *Aegiops kotschyi* chromosome $2S^{k}$ (2R) was carried out using *in situ* hybridisation [4]. Plants carrying the robertsonian translocation were also checked using molecular markers for the presence of the *Lr54* and *Yr37* leaf rust and stripe rust resistance genes. It is clear that *in situ* hybridisation is a cytological procedure that allows the identification of alien chromatin in a given species and it is widely used in breeding programs based on marker-assisted selection.

Introgression lines carrying univalent chromosomes can also be used to generate chromosome rearrangements between the added alien chromosome and the ones from the crop species. Univalent chromosomes can suffer misdivission, chromosome breaks or interspecific recombination with chromosomes from the crop species during meiosis and therefore, series of introgression lines including deleted chromosomes of an alien species can be developed to contribute to the localization of desirable genes. This has been the methodology used to obtain rye introgressions in wheat carrying several deletions of chromosome $5R^{ku}$ from *S. cereale* L. Kustro [5]. *in situ* hybridisation and molecular markers allowed the characterisation of the different deletions of $5R^{ku}$ chromosome and the stripe rust resistant genes location to a region of the long arm of $5R^{ku}$ chromosome.

Fluorescence *in situ* hybridisation has also been used to visualize the chromosomal localization of repetitive sequences that might contribute to chromosome identification. For example, the identification of a novel mini-satellite repeat (Ta-3A1) in wheat has allowed a phylogenetic analysis between wheat and some related species [6]. Non-denaturing *in situ* hybridisation (ND-FISH) was conducted using the Ta-3A1 sequence as a probe on chromosome preparations of wheat and in a long number of related wild and cultivated species, included in the tribe *Triticeae dumort*, as well as in other wheat-alien species amphiploids. The combination of cytogenetics tools and genomic research on repetitive sequences, including the mini-satellite, can contribute to wheat breeding activities, including chromosome manipulation and engineering.

The stability all these chromosome rearrangements are dependent on interspecific chromosome associations and recombination events, which are required during meiosis between the alien chromosomes and those from the crop, to obtain stable chromosome introgressions carrying useful agronomic traits. Thus, the study of chromosome associations during meiosis in general, and chromosome recombination in particular, is of great interest in the framework of chromosome manipulation for plant breeding purposes. A deep review about meiotic recombination within the *Triticeae* tribe, and especially those features concerning polyploidy wheat, is included in this Special Issue. The significance of chromosome rearrangements on meiotic recombination and the importance of recombination in the framework of chromosome manipulation for breeding purposes is also reviewed [7].

3. Chromosome Engineering in other Plant Species

A chromosomal study of other crop species such as coffee (*Coffea arabica* L.) and cacao (*Theobroma cacao* L.), which are of great importance especially for several national economies in Africa, Latin America and Asia, has been included in this Special Issue. Papaya is also a valuable tropical crop for its high nutritional value. Minor tropical crops such as coffee, cacao and papaya do not easily benefit from public and private breeding programs, in contrast to the huge amount of money dedicated to improving major crops, such as wheat. In addition, coffee and cacao are suffering the effect of climate change, accelerating the loss of genetic diversity for future plant breeding programs. Thus, better adapted crops are necessary to facilitate their adaptation to climate change and recombination plays a key role in conventional breeding programs to improve modern varieties by increasing their variability using related species as genetic donors [8]. Other possibilities for chromosome engineering to increase recombination and crossover formation to facilitate the introgression of desirable traits in these tropical crops are also reviewed [8].

Meiwa kumquat (*Fortunella crassifolia* Swingle) is a lesser-known species of kumquat, which was originally classified as *Citrus japonica*, citrus of Japan, until received its own unique genus distinction. A combination of flow cytometry and a microscopy analysis has been performed to clarify the ploidy level of a $2x + 4x$ ploidy chimera of *F. crassifolia* which can be used for breeding purposes [9]. Authors also pointed that this chimera can be used for triploid breeding, where seedless fruits can be expected.

Funding: Experimental work in the Crops Biotechnology Group of P.P. is supported by the Spanish Ministerio de Ciencia e Innovación and The European Regional Development Fund (FEDER) from the European Union.

Conflicts of Interest: The author declares no conflict of interest.

References

1. Palomino, C.; Cabrera, A. Development of wheat-*Hordeum chilense* chromosome 2H^{ch} introgression lines potentially useful for improving grain quality traits. *Agronomy* **2019**, *9*, 493. [CrossRef]
2. Álvarez, J.B.; Castellano, L.; Recio, R.; Cabrera, A. Wx gene in Hordeum chilense: Chromosomal location and characterisation of the allelic variation in the two main ecotypes of the species. *Agronomy* **2019**, *9*, 261. [CrossRef]
3. Kirov, I.; Pirsikov, A.; Milyukova, N.; Dudnikov, M.; Kolenkov, M.; Gruzdev, I.; Siksin, S.; Khrustaleva, L.; Karlov, G.; Soloviev, A. Analysis of wheat bread-making gene (wbm) evolution and occurrence in triticale collection reveal origin via interspecific introgression into chromosome 7AL. *Agronomy* **2019**, *9*, 854. [CrossRef]
4. Ulaszewski, W.; Belter, J.; Wisniewska, A.; Szymczak, J.; Skowronska, J.; Phillips, D.; Kwiatek, M.T. Recovery of 2R.2Sk Triticale-Aegilops kotschyi robertsonian chromosome translocations. *Agronomy* **2019**, *9*, 646. [CrossRef]
5. Xi, W.; Tang, Z.; Luo, J.; Fu, S. Physical location of new stripe rust resistance gene(s) and PCR-based markers on rye (*Secale cereale* L.) chromosome 5 using 5R dissection lines. *Agronomy* **2019**, *9*, 498. [CrossRef]
6. Lang, T.; Li, G.; Yu, Z.; Ma, J.; Chen, Q.; Yang, E.; Yang, Z. Genome-wide distribution of novel Ta-3A1 mini-satellite repeats and its use for chromosome identification in wheat and related species. *Agronomy* **2019**, *9*, 60. [CrossRef]
7. Naranjo, T. The effect of chromosome structure upon meiotic homologous and homoeologous recombinations in Triticeae. *Agronomy* **2019**, *9*, 552. [CrossRef]
8. Bolaños-Villegas, P. Chromosome engineering in tropical cash crops. *Agronomy* **2020**, *10*, 122. [CrossRef]
9. Nukaya, T.; Sudo, M.; Yahata, M.; Ohta, T.; Tominaga, A.; Mukai, H.; Yasuda, K.; Kunitake, H. The confirmation of a ploidy periclinal chimera of the Meiwa Kumquat (*Fortunella crassifolia* Swingle) induced by colchicine treatment to nucellar embryos and its morphological characteristics. *Agronomy* **2019**, *9*, 562. [CrossRef]

Article

Development of wheat—*Hordeum chilense* Chromosome $2H^{ch}$ Introgression Lines Potentially Useful for Improving Grain Quality Traits

Carmen Palomino and Adoracion Cabrera *

Departamento de Genética, Escuela Técnica Superior de Ingeniería Agronómica y de Montes, Edificio Gregor Mendel, Campus de Rabanales, Universidad de Córdoba, CeiA3, ES-14071 Córdoba, Spain
* Correspondence: acabrera@uco.es

Received: 25 July 2019; Accepted: 23 August 2019; Published: 28 August 2019

Abstract: The chromosome $2H^{ch}$ of *Hordeum chilense.* has the potential to improve seed carotenoid content in wheat as it carries a set of endosperm carotenoid-related genes. We have obtained structural changes in chromosome $2H^{ch}$ in a common wheat (*Triticum aestivum* L. "Chinese Spring") background by crossing a wheat double disomic substitution $2H^{ch}$(2D) and $7H^{ch}$(7D) line with a disomic addition line carrying chromosome $2C^{c}$ from *Aegilops cylindrica* Host.. Seven introgressions of chromosome $2H^{ch}$ into wheat were characterized by fluorescence in situ hybridization (FISH) and DNA markers. Chromosome-specific simple sequence repeats (SSRs) were used for identifying wheat chromosomes. In addition, we tested 82 conserved orthologous set (COS) markers for homoeologous group 2, of which 65 amplified targets in *H. chilense* and 26 showed polymorphism between *H. chilense* and wheat. A total of 24 markers were assigned to chromosome $2H^{ch}$ with eight allocated to $2H^{ch}$S and 16 to $2H^{ch}$L. Among the seven introgressions there was a disomic substitution line $2H^{ch}$(2D), a ditelosomic addition line for the $2H^{ch}$L arm, an isochromosome for the $2H^{ch}$L arm, a homozygous centromeric $2H^{ch}$S·2DL translocation, a double monosomic $2H^{ch}$S·2DL plus $7H^{ch}$S·D translocation, a homozygous centromeric $7H^{ch}$S·$2H^{ch}$L translocation and, finally, a $2H^{ch}$L·$7H^{ch}$L translocation. Wheat—*H. chilense* macrosyntenic comparisons using COS markers revealed that *H. chilense* chromosome $2H^{ch}$ exhibits synteny to wheat homoeologous group 2 chromosomes, and the COS markers assigned to this chromosome will facilitate alien gene introgression into wheat. The genetic stocks developed here include new wheat—*H. chilense* recombinations which are useful for studying the effect of chromosome $2H^{ch}$ on grain quality traits.

Keywords: grain colour; *Hordeum chilense*; wheat introgression; wheat quality; wild barley

1. Introduction

Narrow genetic diversity often limits the improvement of many traits in wheat. The introgression of genes from wild relatives to wheat has become a widely recognized genetic approach for increasing genetic diversity and, hence, the need to explore primary, secondary and tertiary gene pools of wheat has grown [1,2]. *Hordeum chilense* Roem. et Schultz. is a diploid wild barley that exhibits advantageous agronomic and quality characteristics [3–6]. Furthermore, its high crossability with other species of the tribe Triticeae, such as both durum and common wheat, [3–5] make it useful in cereal breeding.

Addition and substitution lines of alien chromosomes in a common wheat background, are useful for introgressing alien chromosomal segments carrying genes of agronomical interest into wheat. Chromosome addition lines of *H. chilense* in the *Triticum aestivum* L. cultivar "Chinese Spring" have been obtained including five for chromosomes $1H^{ch}$, $4H^{ch}$, $5H^{ch}$, $6H^{ch}$ and $7H^{ch}$ and the ditelosomic addition line for the short arm of chromosome $2H^{ch}$ [7]. Fertile wheat lines carrying deletions and translocations involving chromosome $3H^{ch}$ from *H. chilense* have also been obtained [8]. However, no addition or substitution lines in a common wheat background have been developed for chromosome $2H^{ch}$.

The location of agronomic traits on specific *H. chilense* chromosomes have been carried using these available wheat—*H. chilense* addition and substitution lines, such as resistance to greenbug (*Schizaphis graminum* Rond.) [9] and endosperm prolamins located on chromosome $1H^{ch}$ [10,11]; resistance to *Septoria tritici* on chromosome $4H^{ch}$ [12]; tolerance to salt on chromosomes $1H^{ch}$, $4H^{ch}$ and $5H^{ch}$ [13]; fertility restoration on chromosome $6H^{ch}$ [14]; and carotenoid content on chromosome $7H^{ch}$ [15]. Wheat—*H. chilense* translocation or recombinant lines have also been generated using both addition and substitution lines of *H. chilense* chromosomes in wheat background [8,11,16–18].

Chromosome $2H^{ch}$ has the potential to improve seed carotenoid content in wheat. Genetic studies of yellow pigment content (YPC) in *H. chilense* revealed that chromosome $2H^{ch}$ showed a significant association with YPC [19] and four endosperm carotenoid-related genes have been genetically mapped to chromosome $2H^{ch}$, such as geranyl geranyl pyrophosphate synthase (*Ggpps1*) for geranylgeranyl diphosphate synthesis, zeta-carotene desaturase (*Zds*), beta-carotene hydroxylase 3 (*Hyd3*) from the carotenoid biosynthetic pathway and polyphenol oxidase 1 gene (*Ppo1*) implicated in plant tissue enzymatic browning [20].

Molecular markers that are able to distinguish *H. chilense* chromosome $2H^{ch}$ in wheat background provide a useful tool for selection. The conserved orthologous set (COS) [21] represents an important reservoir of markers that allow comparative studies with wheat and barley and their transference to *H. chilense* is a main goal.

The aims of this work were the following: (a) to obtain wheat—*H. chilense* chromosome $2H^{ch}$ introgression lines; (b) to characterize the lines obtained by fluorescence *in situ* hybridization (FISH) and chromosome-specific simple sequence repeat (SSR) markers; (c) to transfer COS markers to *H. chilense* and to determine their arm location within $2H^{ch}$ and (d) to compare the arm location with wheat and barley homoeologous group 2.

2. Materials and Methods

2.1. Plant Material

A "Chinese Spring" (CS) wheat—*H. chilense* double $2H^{ch}$(2D)-$7H^{ch}$(7D) disomic substitution line, previously obtained at the University of Córdoba (results not shown), was used for inducing structural changes in chromosome $2H^{ch}$ using gametocidal chromosome $2C^c$ from *Aegilops cylindrica* host. The double $2H^{ch}$(2D) and $7H^{ch}$(7D) disomic substitution line was obtained by pollinating tritordeum (the fertile amphiploid between *H. chilense* and *T. turgidum* L., $AABBH^{ch}H^{ch}$, $2n = 6x = 42$) with a wheat disomic addition line for gametocidal chromosome $2C^c$ from *Ae. cylindrica* Host. following the breeding procedure described in [8]. The double substitution $2H^{ch}$(2D) and $7H^{ch}$(7D) line was pollinated with the wheat disomic addition line for the gametocidal chromosome $2C^c$ from *Ae. cylindrica*. The F1 plants monosomic for $2H^{ch}$, $7H^{ch}$ and $2C^c$ were selfed for four generations.

2.2. Fluorescence In Situ Hybridization (FISH)

The excised root tips were pretreated with ice water for 24 h and then fixed in acetic ethanol: acetic acid (3:1, v/v), as described previously [8]. The FISH protocol was carried out as described by [22]. The pAs1 sequence (1 kb) isolated from *Aegilops tauschii* Coss. [23] and *H. chilense* genomic DNA were used as probes. The pAs1 probe hybridizes to D-genome chromosomes of wheat [24] and H^{ch}-genome chromosomes from *H. chilense* [25]. The pAs1 probe and *H. chilense* DNA were labeled with biotin-16-dUTP (Roche Diagnostics, Switzerland) and with digoxigenin-11-dUTP (Roche Diagnostics, Switzerland), respectively, by nick translation. Three plants per each introgression line were analyzed.

Biotin- and digoxigenin-labelled probes were detected with streptavidin-Cy3 conjugates (Sigma, St. Louis, MO, USA) and antidigoxigenin FITC (Roche Diagnostics) antibodies, respectively. The chromosomes were counterstained with DAPI (4′,6-diamidino-2-phenylindole) and mounted in Vectashield mounting medium (Vector laboratories, Inc., Burlingame, CA, USA). A Leica DMR epifluorescence microscope was used for signal visualization. Images were captured with a Leica DFC7000T camera and processed with LEICA application suite v4.0 software (Leica, Germany).

2.3. Molecular Marker Analysis

A total of 82 COS markers from wheat homoeologous group 2 [21] were studied for their utility in *H. chilense* (File S1). *H. chilense* (line H7) and common wheat CS were used as controls. The CTAB method [26] was used for DNA extraction of young leaf tissue. The concentration of each sample was estimated using a NanoDrop 1000 Spectrophotometer (Thermo Scientific, Waltham, MA, USA). Amplifications were made using a TGradient thermocycler (Biometra, Göttingen, Germany) with 60 ng of template DNA in a 25 µl volume reaction containing 5 µl of 10× PCR Buffer, 0.5 µM of each primer, 1.5–2.0 mM $MgCl_2$, 0.3 mM dNTPs and 0.25 U of Taq DNA polymerase (BIOTOOLS B&M Laboratories, Madrid, Spain). The PCR conditions of COS markers were as follows: 4 min at 94°C, followed by 35 cycles of 45 s at 94°C, 50s at 58°C annealing temperature, 50 s at 72°C, and a final extension step of 7 min at 72 °C.

In addition, four chromosome-specific SSR markers for the wheat D-genome were used for molecular characterization of the introgression lines [27,28]. *Xgwm261* and *Xgwm157* markers were used to detect 2DS and 2DL chromosome arms, respectively. *Xcfd66* and *Xbarc111* were used to detect 7DS and 7DL chromosome arms, respectively. Amplifications were carried out as described at GrainGenes [29] One plant from each introgression line was used for the molecular characterization. "Chinese Spring", *H. chilense*, a ditelosomic $2H^{ch}S$ line, a ditelosomic $7H^{ch}S$ line, a ditelosomic $7H^{ch}L$ line and disomic substitution line CS $7H^{ch}$(7D) were used as controls. Ditelosomic $2H^{ch}S$ and ditelosomic $7H^{ch}S$ lines were provided by the John Innes Centre (UK). The ditelosomic $7H^{ch}L$ line and disomic CS $7H^{ch}$(7D) substitution line were obtained previously [17].

The amplified products were resolved using 2% agarose gels (SSRs) or polyacrylamide gels (10%, w/v; C: 2.67%) (COS) and stained with ethidium bromide or SafeView Nucleic Acid Stain (NBS Biologicals, Huntingdon, UK) incorporated in the gel. A 100 bp DNA ladder (Solis BioDyne, Tartu, Estonia) was used as a standard molecular weight marker. Kodak Digital Science 1D software (version 2.0) was used to determine the amplicon lengths.

2.4. Comparative Mapping

The orthologous relationship between the 2A, 2B, and 2D genome chromosomes of bread wheat and the $2H^{ch}$ chromosome from *H. chilense* has been studied from the genomic perspective of wheat as described previously [30]. For the construction of the physical map, the expressed sequence tag (EST) source sequences (File S2) were used as queries in BLASTn searches against the wheat reference pseudomolecules [31] to identify the start positions (bp) of the ESTs. In this study, BLAST hits with E values smaller than $1e^{-10}$, identity % > 58.44 and alignment length > 100 bp were considered significant. The genomic start positions in bp of the best hits in wheat pseudomolecules (File S3) were used to construct a physical map of the polymorphic COS markers. The wheat reference genome sequence [31] was used to determine the centromere positions for 2A, 2B and 2D wheat chromosomes. Both the length in bp of wheat pseudomolecules, as well as the start genomic positions of the ESTs, were converted to pixels. Then, the data from the BLASTn searches were used to construct a physical map for 2A, 2B, and 2D wheat chromosomes showing the position of the source EST of the COS markers assigned to *H. chilense* chromosome $2H^{ch}$.

The rice locus (RAP) [21] was used to locate the COS markers in the barley genome zipper [32]. The RAP locus identifier was retrieved using the ID Converter tool [33]. The full-length barley cDNA corresponding to each rice locus was used for determination of the barley Unigene corresponding to each COS marker. The Unigene sequences were aligned in Barleymap [34] to obtain their positions in the International Barley Sequencing Consortium map [32,35].

3. Results

3.1. Cytogenetic and Molecular Characterization of Wheat—*H. chilense* Introgression Lines Involving Chromosome $2H^{ch}$

The pAs1 and *H. chilense* genomic DNA used as probes in FISH analysis allowed the identification of a pair of $2H^{ch}$ chromosomes and the absence of the wheat 2D chromosome pair in one line with 42 chromosomes. This result indicated that this line was disomic for the substitution $2H^{ch}$(2D) (Figure 1a). The absence of 2D was tested using *Xgwm261-2DS* (Figure 2a) and *Xgwm157-2DL* (Figure 2b) molecular markers. A pair of telocentric chromosomes was identified by FISH in one line with 42 + 2t chromosomes (Figure 1b). To determine the chromosome arm involved in each introgression line, we used the *c749557* COS marker mapped on the $2H^{ch}S$ arm and *c731690* mapped on $2H^{ch}L$, respectively (see Section 3.2). The presence of the c731690 marker for $2H^{ch}L$ and the absence of the *c749557* marker for $2H^{ch}S$ showed that this line was ditelosomic for the $2H^{ch}L$ arm (Figure 3a,b).

FISH analysis revealed a line apparently carrying chromosome $2H^{ch}$ (Figure 1c). Marker *c731690* for $2H^{ch}L$ was amplified in this line, but there was no amplification of the *c749557* marker for $2H^{ch}S$ (Figure 3a,b). These results suggested that a $2H^{ch}L{\cdot}2H^{ch}L$ isochromosome was present in this line, and it was named Iso $2H^{ch}L$. Both the ditelosomic $2H^{ch}L$ and Iso $2H^{ch}L$ lines were nullisomic for chromosome 2D, as demonstrated by the absence of amplification of both *Xgwm261-2DS* (Figure 2a) and *Xgwm157-2DL* (Figure 2b) molecular markers.

We identified two lines carrying centromeric translocations involving the $2H^{ch}S$ chromosome arm and wheat chromosomes. One of these lines was homozygous for the $2H^{ch}S{\cdot}2DL$ translocation (Figure 1d). The other translocation line was a double monosomic for $2H^{ch}S{\cdot}2DL$ and $7H^{ch}S{\cdot}D$ translocations (Figure 1e). Chromosome-specific SSR markers confirmed the absence of 2DS (Figure 2a) and the presence of 2DL (Figure 2b) in both lines. Amplification of the *c749557* marker (Figure 3a) and the absence of amplification of the *c731690* marker (Figure 3b) demonstrated the presence of $2H^{ch}S$ and the absence of $2H^{ch}L$, respectively, in both lines. COS markers *c779791* and *c759439*, previously assigned to $7H^{ch}S$ and $7H^{ch}L$, respectively [18], were used to detect introgression from chromosome $7H^{ch}$ (Figure 3c,d). Amplification of the *c779791* marker specific for the $7H^{ch}S$ arm (Figure 3c) indicated the presence of $7H^{ch}S$ translocated to an unidentified wheat fragment. The presence of pAs1 signals on the wheat small fragment indicated that the chromosome $7H^{ch}S$ arm was translocated to an unidentified D-genome chromosome (Figure 1e). The absence of amplification of chromosome-specific markers *Xcfd66*-7DS and *Xbarc111*-7DL demonstrated the absence of a 7D chromosome pair in this line (Figure 2c,d).

In the remaining two lines, two centromeric translocations involving $2H^{ch}$ and $7H^{ch}$ *H. chilense* chromosomes were detected. One line was homozygous for the $7H^{ch}S{\cdot}2H^{ch}L$ translocation (Figure 1f) and the other one was monosomic for the $2H^{ch}L{\cdot}7H^{ch}L$ translocation. Both translocation lines were nullisomic for chromosome 2D (Figure 2a,b). Chromosome-specific SSR marker patterns for 2D (*Xgwm261-2DS* and *Xgwm157-2DL*) and 7D (*Xcfd66-7DS* and *Xbarc111-7DL*) genome chromosomes are given in Figure 2a,b and Figure 2c,d, respectively. Chromosome-specific marker results for chromosome $7H^{ch}S$ and $7H^{ch}L$ are given in Figure 3c,d. Table 1 shows the chromosome constitutions of all the *H. chilense* introgression lines. All lines were vigorous and seed set.

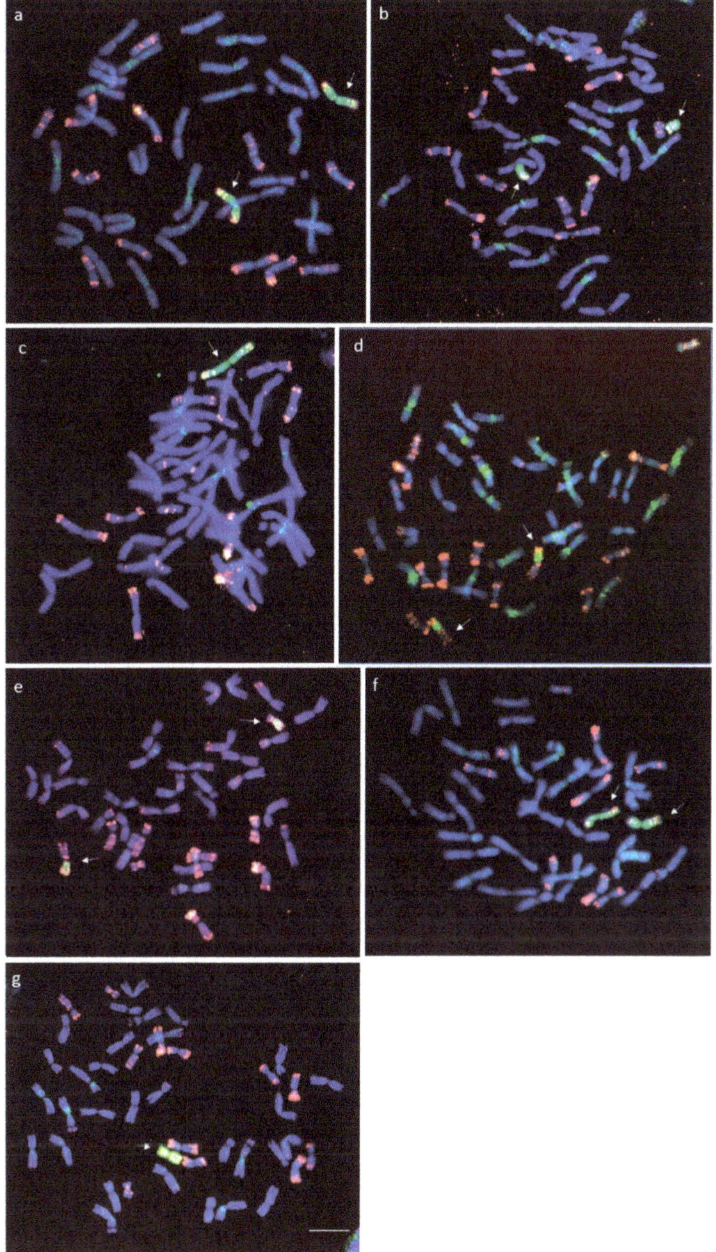

Figure 1. Fluorescence in situ hybridization (FISH) with the pAs1 repetitive (red) and *H. chilense* genomic DNA (green) probes to mitotic metaphase of wheat—*H. chilense* introgression lines involving chromosome $2H^{ch}$. (**a**) Disomic substitution $2H^{ch}$ (2D); (**b**) Ditelosomic $2H^{ch}L$; (**c**) Isochromosome $2H^{ch}L$; (**d**) Translocation $2H^{ch}S{\cdot}2DL$; (**e**) Translocation $2H^{ch}S{\cdot}2DL$ + $T7H^{ch}S{\cdot}D$; (**f**) Translocation $7H^{ch}S{\cdot}2H^{ch}L$; (**g**) Translocation $2H^{ch}L{\cdot}7H^{ch}L$. Bar = 10 μm.

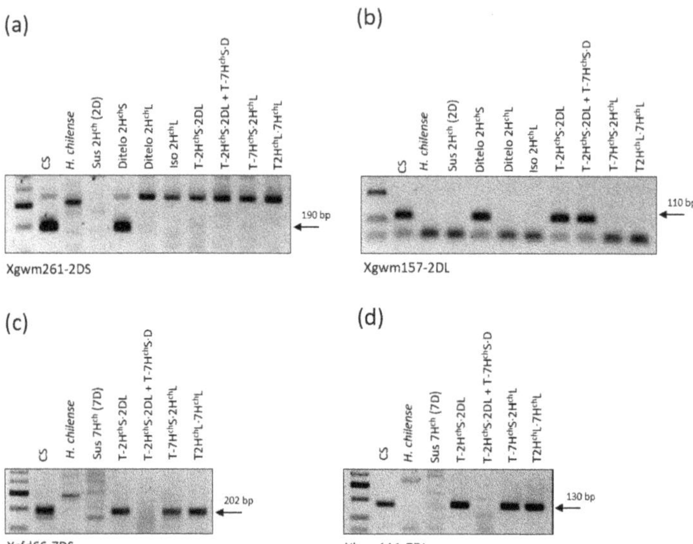

Figure 2. Molecular characterization of introgression lines with wheat chromosome-specific simple sequence repeats (SSR) markers. (**a**) *Xgwm261*-2DS; (**b**) *Xgwm157*-2DL; (**c**) *Xcfd66*-7DS and (**d**) *Xbarc111*-7DL.

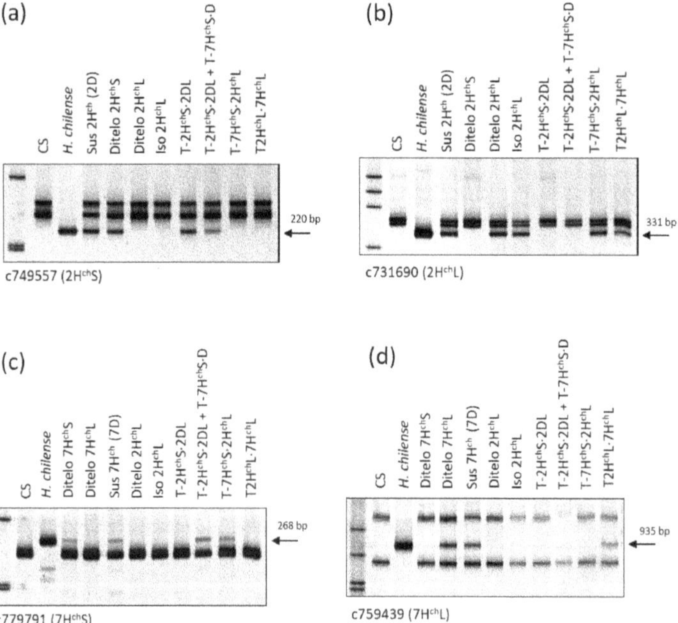

Figure 3. Examples of PCR amplification profiles used for identifying chromosome 2H^{ch} and 7H^{ch} arms in the introgression lines. (**a**) c749557 mapped on the short arm of chromosome 2H^{ch}; (**b**) c731690 mapped on the long arm of chromosome 2H^{ch}; (**c**) c779791 mapped on the short arm of chromosome 7H^{ch}; and (**d**) c759439 mapped on the long arm of chromosome 7H^{ch}. "Chinese Spring" (CS), *H. chilense* (H^{ch}), ditelo 2H^{ch}S, ditelo 7H^{ch}S, ditelo 7H^{ch}L and disomic substitution line CS 7H^{ch}(7D) were used as controls.

Table 1. Chromosome constitutions of wheat—*H. chilense* introgression lines involving chromosome $2H^{ch}$.

Line	Type of Aberration	*H. chilense* Introgressions	No. of D Chromosomes	No. of A/B Chromosomes	Total No. of Chromosomes
Sus $2H^{ch}$ (2D)	Substitution [1]	$2H^{ch}$	12 (2D pair absent)	28	42
Ditelo $2H^{ch}L$	Telosome [1]	$2H^{ch}L$	12 (2D pair absent)	30	42 + 2 telos
Iso $2H^{ch}L$	Isochromosome [2]	$2H^{ch}L$	12 (2D pair absent)	30	42 + iso
$T2H^{ch}S{\cdot}2DL$	Translocation [1]	$2H^{ch}S$	12 + 2T (2DL absent)	30	42 + 2T
$T2H^{ch}S{\cdot}2DL$ + $T7H^{ch}S{\cdot}D$	Translocation [3]	$2H^{ch}S+7H^{ch}S$	8 + 2T (7D pair absent)	28	36 + 2T
$T7H^{ch}S{\cdot}2H^{ch}L$	Translocation [1]	$2H^{ch}L+7H^{ch}S$	9 (2D pair absent)	30	39 + 2T
$T2H^{ch}L{\cdot}7H^{ch}L$	Translocation [2]	$2H^{ch}L+7H^{ch}L$	12 (2D pair absent)	30	42 + 1T

[1] disomic; [2] monosomic; [3] double monosomic

3.2. Transferability and Chromosome Location of COS Markers in H. chilense

The transferability to *H. chilense* of 83 COS markers from wheat homoeologous group 2 was studied (File S1). First, all 83 markers were screened for polymorphisms (size polymorphisms or presence and absence) between *H. chilense* and common wheat. Of the 83 markers, 65 (78.3%) consistently amplified *H. chilense* products and 26 (40.0% of the total) were polymorphic between *H. chilense* and wheat (Table 2). Twenty-four of these 26 polymorphic markers were mapped to chromosome $2H^{ch}$, as demonstrated by their presence in the wheat—*H. chilense* $2H^{ch}$(2D) substitution line. We were unable to map the remaining two markers because they did not amplify products in any of the available wheat—*H. chilense* addition lines. Of the 24 COS markers mapped on chromosome $2H^{ch}$, eight were located on $2H^{ch}S$ and 16 were located on $2H^{ch}L$, as demonstrated by their presence and absence in $2H^{ch}S$ or $2H^{ch}L$ ditelosomic lines, respectively. Table 2 summarize the characterization and chromosome arm location of wheat COS markers on *H. chilense* chromosome $2H^{ch}$. Figure 3a,b shows examples of amplification of homoeologous group 2 COS markers.

Table 2. Characterization and chromosome localization of wheat conserved orthologous set (COS) markers on *H. chilense* chromosome $2H^{ch}$.

Marker	Product Size in *T. aestivum*	Product Size in *H. chilense*	Arm Location in *H. chilense*	Chromosome Location in Wheat [1]	Location in Wheat (cm) [1]	Location in Barley (cm) [2]
c723421	262–234	242	$2H^{ch}S$	2BS	28.1	63.5
c754613	775	750	$2H^{ch}S$	2AS-2BS-2DS	31.7	46.3
c745448	329	313–364	$2H^{ch}S$	2BS	32.1	52.5
c77095	997–1075	1186	$2H^{ch}S$	2AS-2BS-2DS	35.9	47.7
c741602	887	925	$2H^{ch}S$	2AS-2BS	45.8	56.3
c751379	889	911	$2H^{ch}S$	2AS-2BS	56.3	59.2
c733078	419	393	$2H^{ch}S$	2BS-2DS	68.2	Not found
c756234	432	311	-	2AS-2BS	63.9	52.2
c749557	251–268	220	$2H^{ch}S$	2AS-2DS	69.0	59.2
c740970	229	239	$2H^{ch}L$	2AL-2BL-2DL	140.2	63.5
c744070	204	221	$2H^{ch}L$	2AL-2BL-2DL	143.7	Not found
c729382	602	463	$2H^{ch}L$	2BL-2DL	145.5	69.2
c751189	805	851	$2H^{ch}L$	2AL-2BL-2DL	145.8	70.54
c749297	257	247	$2H^{ch}L$	2AL-2BL-2DL	149.3	71.1
BF291656	920	821	$2H^{ch}L$	2AL-2BL-2DL	152.0	98.6
c760814	554602		$2H^{ch}L$	2DL	167.4	Not found
c747571	330	370	$2H^{ch}L$	2AL-2BL-2DL	173.2	122.2
c743473	252	276	$2H^{ch}L$	2AL-2BL-2DL	173.5	82.75
c779794	521–602	661	$2H^{ch}L$	2AL-2BL	176.0	113.5
c795564	573	629	$2H^{ch}L$	2AL	178.5	113.5

Table 2. *Cont.*

Marker	Product Size in *T. aestivum*	Product Size in *H. chilense*	Arm Location in *H. chilense*	Chromosome Location in Wheat [1]	Location in Wheat (cm) [1]	Location in Barley (cm) [2]
c730704	377	393	$2H^{ch}L$	2DL	179.4	156.7
c741642	366-390	377	$2H^{ch}L$	2BL-2DL	187.3	126.0
c746642	843–889	864	$2H^{ch}L$	2AL-2BL-2DL	189.5	156.7
c747195	508	668	$2H^{ch}L$	2AL-2BL-2DL	190.2	Not found
c731690	342–355	320	$2H^{ch}L$	2BL-2DL	191.7	102.8
c756123	398–520	442	-	2AL-2BL-2DL	198.2	136.8

[1] Quraishi et al. (2009); [2] Position in barley determined using Barleymap (http://floresta.eead.csic.es).

3.3. Wheat—*H. chilense* Group 2 Homoeology

To investigate wheat—*H. chilense* group 2 macrosyntenic relationships, the source ESTs of the 24 polymorphic COS markers mapped on chromosome $2H^{ch}$ were BLASTed to the sequences of the wheat chromosomes [31]. All EST markers showed hits on wheat pseudomolecules (File S3). To produce a physical map (Figure 4), the start positions of the alignments of the best hits on the A, B, and D genomes were extracted. All markers assigned to $2H^{ch}S$ and $2H^{ch}L$ were located in the same arm of wheat homoeologous group 2 chromosomes (Table 2, Figure 4).

To determine the positions of COS markers in Barley-maps (Table 2) the align tool implemented in Barleymap [34] was used to determine the positions of COS markers in barley maps (Table 2). Four of the COS markers were not found in the barley genome zipper [32] so their position in barley could not be determined. A good correspondence between *H. chilense* and barley was found for the arm locations of COS markers (Table 2).

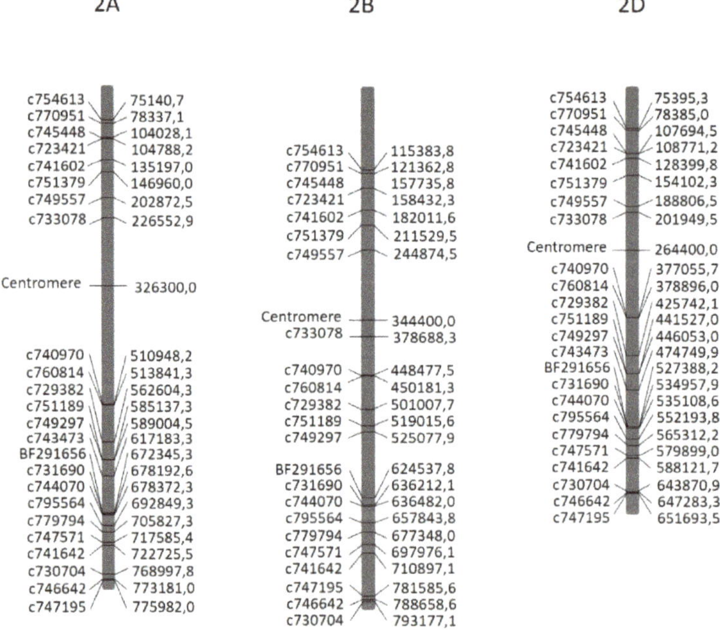

Figure 4. Visualization of wheat-*H. chilense* orthologous regions from the perspective of wheat homoeologous chromosome group 2. Physical map of the source expressed sequence tags (ESTs) of the COS-markers (left), the genomic positions on wheat pseudomolecules (kb) are on the right.

4. Discussion

A set of introgression lines involving chromosome $2H^{ch}$ from *H. chilense* in common wheat background was produced in this work using the gametocidal chromosome $2C^{c}$ from *Ae. cylindrica*. Gametocidal genes have been used to produce structural chromosome aberrations in wheat [36,37], barley [38,39] and rye [40]. In *H. chilense*, structural changes have been previously obtained for chromosomes $1H^{ch}$, $3H^{ch}$, $4H^{ch}$ and $7H^{ch}$ [8,11,17,18,41] and have been useful in determining the locations of genes and markers in this species. Breaks at both centromeric and interstitial regions of chromosomes have been induced by gametocidal chromosome $2C^{c}$ [36]. In the present study, telocentric and translocations between chromosome $2H^{ch}$ from *H. chilense* and wheat chromosomes have been generated

Alien addition and translocation lines are an ideal template for PCR-based mapping to assign molecular markers to chromosomes of the wild relatives of wheat [8,11,41–43]. Using gene-based conserved orthologous set (COS) markers on wheat—*H. chilense* introgression lines obtained in this work, we assigned a total of 24 markers to *H. chilense* chromosome $2H^{ch}$. A 78.3% transference rate of COS markers to *H. chilense* chromosome $2H^{ch}$ was found. Since COS markers were intended for comparative studies among grasses, the high rate of transferability obtained in this work was expected [21]. Similar rates of transference of COS markers to *H. chilense* chromosome $7H^{ch}$ have been found previously [17]. COS markers have also been transferred successfully to other Triticeae species such as *Agropyron cristatum* [29,41] and *Aegilops* spp. [44].

The relevance of chromosome $2H^{ch}$ for endosperm carotenoid content has been highlighted by previous work. Association studies for YPC allowed the identification of three main chromosome regions for YPC variation in *H. chilense*, with the largest one located on chromosome $2H^{ch}$ and smaller regions detected on chromosomes $3H^{ch}$ and $7H^{ch}$ [20]. Four candidate genes associated with YPC were genetically mapped to chromosome $2H^{ch}$: both *Ggpps1* and *Zds* were tightly linked and mapped near the centromere, while *Hyd3* and *Ppo1* were mapped to the long arm of chromosome $2H^{ch}$ [20]. Furthermore, a significant QTL at the distal part of chromosome $2H^{ch}$ has also been found where no carotenoid-related genes have been mapped [19,20]. The importance of chromosome $2H^{ch}L$ in grain carotenoid content has also been revealed in the new cereal tritordeum (amphiploid derived from a cross between the wild barley *H. chilense* and durum wheat), which has higher carotenoid pigment content in its grain than durum or bread wheat [45,46].

Chromosome $7H^{ch}$ from *H. chilense* confers the capacity to accumulate higher carotene concentration in seeds [15]. The *Psy1* gene controlling the first step of the carotenoid biosynthetic pathway was mapped to $7H^{ch}S$ [47]. Wheat—*H. chilense* chromosome $7H^{ch}$introgression lines have been developed [17,18,48], and all the genetic stocks carrying *Psy1* from *H. chilense* show increased carotenoid content relative to common wheat [48,49]. The obtention in this work of translocation $T7H^{ch}S{\cdot}2H^{ch}L$ could be of interest for studying the effect of the *Psy1* gene located on the $7H^{ch}S$ arm and both the *Hyd3* and *Ppo1* genes mapped to the long arm of chromosome $2H^{ch}$ [20]. Furthermore, the lines carrying $T2H^{ch}S{\cdot}2DL$ and $T2H^{ch}S{\cdot}2DL + T7H^{ch}S{\cdot}D$ translocations could be of interest for studying the effect of both the *Ggpps1* and *Zds* genes mapped to the short arm of $2H^{ch}$, in the absence and presence of the *Psy1* gene from *H. chilense*, respectively.

The transference of desirable genes from wild relatives to wheat can be restrict by linkage drag and the lack of compensation for the wheat chromatin substituted. The identification of alien chromosomal regions carrying the genes of interest and the analysis of their homoeologous relationships with wheat chromosomes can overcome that difficulty. It has been pointed out that only translocations produced by homoeologous recombination are beneficial for wheat improvement [50,51]. In this work, wheat—*H. chilense* macrosyntenic comparisons using COS markers revealed that *H. chilense* chromosome $2H^{ch}$ exhibits good synteny with wheat homoeologous group 2 chromosomes. Comparative mapping of carotenoid-related genes mapped to chromosome $2H^{ch}$ also showed good collinearity between *H. chilense* and Triticeae species [20,46]. The 24 COS markers assigned to chromosome $2H^{ch}$ in this work will facilitate introgression of alien genes associated with this chromosome into wheat.

5. Conclusions

We used in situ hybridization and genotyping to characterize genetic stocks harboring chromosome $2H^{ch}$ from *H. chilense* in a wheat background. As far as we know, no previous translocation lines involving chromosome $2H^{ch}$introgressed into wheat have been described. The cytogenetic stocks developed here may constitute an important resource for studying the effect of chromosome $2H^{ch}$ on wheat grain color. In addition, these cytogenetic stocks allowed the localization of a set of conserved orthologous set (COS) markers to specific arms in chromosome $2H^{ch}$.The genomic position of orthologous unigene EST-contigs used for the COS marker design revealed a good macrosyntenic relationship between *H. chilense* chromosome $2H^{ch}$ and wheat homoeologous group 2. The new wheat—*H. chilense* recombinations are useful for genetic studies and might also serve as a bridge for transferring genes associated with yellow pigment into a wheat background.

Supplementary Materials: The following are available online at http://www.mdpi.com/2073-4395/9/9/493/s1, File S1: COS markers used in this work together with their primer sequences, File S2: Source of the COS markers assigned to the *H. chilense* chromosome $2H^{ch}$, File S3: Results of BLASTn search for COS markers assigned to *H. chilense* chromosome $2H^{ch}$ in the reference sequences of hexaploid wheat group 2 chromosomes (www.wheatgenome.org/) and the start positions (bp) of the marker-specific ESTs.

Author Contributions: A.C. conceived and designed the study; C.P. performed the experiments; A.C. analysed the data and wrote the paper: all authors have read and approved the final manuscript.

Funding: This research was supported by grants AGL2014-53195-R and RTI2018-093367-B-I00 from the Spanish State Research Agency (Ministry of Science, Innovation and Universities), co-financed by the European Regional Development Fund (FEDER) from the European Union.

Conflicts of Interest: The authors declare no conflict of interest.

References

1. Hajjar, R.; Hodgkin, T. The use of wild relatives in crop improvement: A survey of developments over the last 20 years. *Euphytica* **2007**, *156*, 13. [CrossRef]
2. Prohems, J.; Gramazio, P.; Plazas, M.; Dempewil, H.; Kilian, B.; Díez, M.J.; Fita, A.; Herraiz, F.J.; Rodríguez-Burruezo, A.; Soler, S.; et al. Introgressiomics: A new approach for using wild relatives in breeding for adaptation to climate change. *Euphytica* **2017**, *213*, 158. [CrossRef]
3. Martín, A.; Martín, L.M.; Cabrera, A.; Ramírez, M.C.; Giménez, M.J.; Rubiales, D.; Hernández, P.; Ballesteros, J. The potential of *Hordeum chilense* in breeding Triticeae species. In *Triticeae III*; Jaradat, A.A., Humphreys, M., Eds.; Science Publishers Inc.: Enfield, UK, 1998; pp. 377–386.
4. Martín, A.; Cabrera, A.; Hernández, P.; Ramírez, M.C.; Rubiales, D.; Ballesteros, J. Prospect for the use of *Hordeum chilense* in durum wheat breeding. In *Durum Wheat Improvement in the Mediterranean Region: New Challenges*; Rojo, C., Nachit, M., Di-Fonzo, N., Araus, J., Eds.; Options Méditerranées: Zaragoza, Spain, 2000; pp. 111–115.
5. Martín, A.; Cabrera, A. Cytogenetics of *Hordeum chilense*: Current status and considerations with reference to breeding. *Cytogenet. Genome Res.* **2005**, *109*, 378–384. [CrossRef] [PubMed]
6. Alvarez, J.B.; Guzman, C. Interspecific and intergeneric hybridization as a source of variation for wheat grain quality improvement. *Theor. Appl. Genet.* **2018**, *131*, 225–251. [CrossRef] [PubMed]
7. Miller, T.E.; Reader, S.M.; Chapman, V. The addition of *Hordeum chilense* chromosomes to wheat. In *Induced Variability in Plant Breeding, Proceedings of International Symposium in Eucarpia*; Broertjes, C., Ed.; Pudoc: Wageningen, The Netherlands, 1982; pp. 79–81.
8. Said, M.; Recio, R.; Cabrera, A. Development and characterisation of structural changes in chromosome $3H^{ch}$ from *Hordeum chilense* in common wheat and their use in physical mapping. *Euphytica* **2012**, *188*, 429–440. [CrossRef]
9. Castro, A.M.; Martin, A.; Martin, L.M. Location of genes controlling resistance to greenbug (*Scizaphis graminum* Rond.) in *Hordeum chilense*. *Plant Breed* **1996**, *115*, 335–338. [CrossRef]
10. Payne, P.I.; Holt, L.M.; Reader, S.M.; Miller, T.E. Chromosomal location of genes coding for endosperm proteins of *Hordeum chilense*, determined by two-dimensional electrophoresis of wheat-*H. chilense* chromosome addition lines. *Biochem. Genet.* **1987**, *25*, 53–65. [CrossRef] [PubMed]

11. Cherif-Mouaki, S.; Said, M.; Alvarez, J.B.; Cabrera, A. Sub-arm location of prolamin and EST-SSR loci on chromosome $1H^{ch}$ from *Hordeum chilense*. *Euphytica* **2011**, *178*, 63–69. [CrossRef]
12. Rubiales, D.; Reader, S.M.; Martín, A. Chromosomal location of resistance of *Septoria tritici* in *Hordeum chilense* determined by the study of chromosomal addition and substitution lines in "Chinese Spring" wheat. *Euphytica* **2000**, *115*, 221–224. [CrossRef]
13. Forster, B.P.; Phillips, M.S.; Miller, T.E.; Baird, E.; Powell, W. Chromosome location of genes controlling tolerance to salt (NaCl) and vigour in *Hordeum vulgare* and *H. chilense*. *Heredity* **1889**, *65*, 9–107. [CrossRef]
14. Martin, A.C.; Atienza, S.G.; Ramírez, M.C.; Barro, F.; Martin, A. Male fertility restoration of wheat in *Hordeum chilense* cytoplasm is associated with $6H^{ch}S$ chromosome addition. *Aust. J. Agric. Res.* **2008**, *59*, 206–213. [CrossRef]
15. Alvarez, J.B.; Martín, L.M.; Martín, A. Chromosomal localization of genes for carotenoid pigments using addition lines of *Hordeum chilense* in wheat. *Plant Breed.* **1998**, *117*, 287–289. [CrossRef]
16. Calderón, M.D.C.; Ramírez, M.D.C.; Martín, A.; Prieto, P. Development of *Hordeum chilense* $4H^{ch}$ introgression lines in durum wheat: A tool for breeders and complex trait analysis. *Plant Breed.* **2012**, *131*, 733–738. [CrossRef]
17. Mattera, M.G.; Avila, M.C.; Atienza, S.G.; Cabrera, A. Cytological and molecular characterization of wheat-*Hordeum chilense* chromosome $7H^{ch}$ introgression lines. *Euphytica* **2015**, *203*, 165–176. [CrossRef]
18. Mattera, M.G.; Cabrera, A. Characterization of a set of common wheat-*Hordeum chilense* chromosome $7H^{ch}$ introgression lines and its potential use in research on grain quality traits. *Plant Breed.* **2017**, *136*, 344–350. [CrossRef]
19. Atienza, S.G.; Ramírez, C.M.; Hernández, P.; Martín, A. Chromosomal location of genes for carotenoid pigments in *Hordeum chilense*. *Plant Breed.* **2004**, *123*, 303–304. [CrossRef]
20. Rodríguez-Suárez, C.; Atienza, S.G. *Hordeum chilense* genome, a useful tool to investigate the endosperm yellow pigment content in the Triticeae. *BMC Plant Biol.* **2012**, *12*, 200. [CrossRef]
21. Quraishi, U.M.; Abrouk, M.; Bolot, S.; Pont, C.; Throude, M.; Guilhot, N.; Confolent, C.; Bortolini, F.; Praud, S.; Murigneux, A.; et al. Genomics in cereals: From genome-wide conserved orthologous set (COS) sequences to candidate genes for trait dissection. *Func. Integr. Genom.* **2009**, *9*, 473–484. [CrossRef]
22. Cabrera, A.; Martín, A.; Barro, F. In-Situ Comparative Mapping (ISCM) of *Glu-1* loci in *Triticum* and *Hordeum*. *Chromosome Res.* **2002**, *10*, 49–54. [CrossRef]
23. Rayburn, A.L.; Gill, B.S. Molecular identification of the D-genome chromosomes of wheat. *J. Hered.* **1986**, *77*, 253–255. [CrossRef]
24. Mukai, Y.; Nakahara, Y.; Yamamoto, M. Simultaneous discrimination of the three genomes in hexaploid wheat by multicolor fluorescence in situ hybridization using total genomic and highly repeated DNA probes. *Genome* **1993**, *36*, 489–494. [CrossRef] [PubMed]
25. Cabrera, A.; Friebe, B.; Jiang, J.; Gill, B.S. Characterization of *Hordeum chilense* chromosomes by C-banding and in situ hybridization using highly repeated DNA probes. *Genome* **1995**, *38*, 435–442. [CrossRef] [PubMed]
26. Murray, M.G.; Thompson, W.F. Rapid isolation of high molecular weight plant DNA. *Nucleic Acid Res.* **1980**, *8*, 4321–4325. [CrossRef] [PubMed]
27. Röder, M.S.; Korzun, V.; Wandehake, K.; Planschke, J.; Tixier, M.H.; Leroy, P.; Ganal, M.W. A microsatellite map of wheat. *Genetics* **1998**, *149*, 2007–2023. [PubMed]
28. Sourdille, P.; Singh, S.; Cadalen, T.; Brown-Guedira, G.; Gay, G.; Qi, L.; Gill, B.S.; Dufour, P.; Murigneux, A.; Bernard, M. Microsatellite-based deletion bin system for the establishment of genetic-physical map relationships in wheat (*Triticum aestivum* L.). *Func. Integr. Genom.* **2004**, *4*, 12–25. [CrossRef] [PubMed]
29. GrainGenes. Available online: http://wheat.pw.usda.gov/cgi-bin/graingenes/browse.cgi?class=marker (accessed on 1 December 2018).
30. Said, M.; Copete, A.; Gaál, E.; Molnár, I.; Cabrera, A.; Doležel, J.; Vrána, J. Uncovering macrosyntenic relationships between tetraploid *Agropyron cristatum* and bread wheat genomes using COS markers. *Theor. Appl. Genet.* **2019**. [CrossRef] [PubMed]
31. International Wheat Genome Sequencing Consortium. Shifting the limits in wheat research and breeding using a fully annotated reference genome. *Science* **2018**, *361*, eaar7191. [CrossRef]
32. Mayer, K.F.X.; Martis, M.; Hedley, P.E.; Simková, H.; Liu, H.; Morris, J.A.; Steuernagel, B.; Taudien, S.; Roessner, S.; Gundlach, H.; et al. Unlocking the barley genome by chromosomal and comparative genomics. *Plant Cell* **2011**, *23*, 1249–1263. [CrossRef]

33. RAP-DB The Rice Annotation Project Database. Available online: https://rapdb.dna.affrc.go.jp/tools/converter (accessed on 22 November 2018).
34. Cantalapiedra, C.P.; Boudiar, R.; Casas, A.M.; Igartua, E.; Contreras-Moreira, B. Barleymap: Physical and genetic mapping of nucleotide sequences and annotation of surrounding loci in barley. *Mol. Breed.* **2014**, *35*, 13. [CrossRef]
35. International Barley Genome Sequencing Consortium. A physical, genetic and functional sequence assembly of the barley genome. *Nature* **2012**, *491*, 711–716. [CrossRef]
36. Endo, T.R. Induction of chromosomal structural changes by a chromosome of *Aegilops cylindrica* L. in common wheat. *J. Hered.* **1988**, *79*, 366–370. [CrossRef]
37. Endo, T.R.; Gill, B.S. The deletion stocks of common wheat. *J. Hered.* **1996**, *87*, 295–307. [CrossRef]
38. Endo, T.R. Cytological dissection of barley genome by the gametocidal system. *Breed. Sci.* **2009**, *59*, 481–486. [CrossRef]
39. Nasuda, S.; Kikkawa, Y.; Ashida, T.; RafiqulIslam, A.K.M.; Sato, K.; Endo, R. Chromosomal assignment and deletion mapping of barley EST markers. *Genes Genet. Syst.* **2005**, *80*, 357–366. [CrossRef] [PubMed]
40. Friebe, B.; Kynast, R.G.; Gill, B.S. Gametocidal factor induced structural rearrangements in rye chromosomes added to common wheat. *Chromosome Res.* **2000**, *8*, 501–511. [CrossRef] [PubMed]
41. Said, M.; Cabrera, A. A physical map of chromosome $4H^{ch}$ from *Hordeum chilense* containing SSR, STS and EST-SSR molecular markers. *Euphytica* **2009**, *167*, 253–259. [CrossRef]
42. Ochoa, V.; Madrid, E.; Said, M.; Rubiales, D.; Cabrera, A. Molecular and cytogenetic characterization of a common wheat-*Agropyron cristatum* chromosome translocation conferring resistance to leaf rust. *Euphytica* **2015**, *201*, 89–95. [CrossRef]
43. Copete, A.; Cabrera, A. Chromosomal location of genes for resistance to powdery mildew in *Agropyron cristatum* and mapping of Conserved Orthologous Set molecular markers. *Euphytica* **2017**, *213*, 189–297. [CrossRef]
44. Molnár, I.; Vrána, J.; Burešová, V.; Cápal, P.; Farkas, A.; Cseh, A.; Kubaláková, M.; Molnár-Láng, M.; Doležel, J. Dissecting the U, M, S and C genomes of wild relatives of bread wheat (*Aegilops* spp.) into chromosomes and exploring their synteny with wheat. *Plant J.* **2016**, *88*, 452–467. [CrossRef]
45. Atienza, S.G.; Ballesteros, J.; Martín, A.; Hornero-Méndez, D. Genetic variability of carotenoid concentration and degree of esterification among tritordeum (xTritordeumAscherson et Graebner) and durum wheat accessions. *J. Agric. Food Chem.* **2007**, *55*, 4244–4251. [CrossRef]
46. Rodríguez-Suárez, C.; Atienza, S.G. Polyphenol oxidase genes in *Hordeum chilense* and implications in tritordeum breeding. *Mol. Breed.* **2014**, *34*, s11032–s11014. [CrossRef]
47. Atienza, S.G.; Ávila, C.M.; Martín, A. The development of a PCR-based marker for PSY1 from *Hordeum chilense*, a candidate gene for carotenoid content accumulation in tritordeum seeds. *Crop Pasture Sci.* **2007**, *58*, 767–773. [CrossRef]
48. Rey, M.D.; Calderón, M.D.C.; Rodrigo, M.J.; Zacarías, L.; Alós, E.; Prieto, P. Novel bread wheat lines enriched in carotenoids carrying *Hordeum chilense* chromosome arms in the *ph1b* background. *PLoS ONE* **2015**, *10*, e0134598. [CrossRef] [PubMed]
49. Mattera, M.G.; Cabrera, A.; Hornero-Méndez, D.; Atienza, S.G. Lutein esterification in wheat endosperm is controlled by the homoeologous group 7 and is increased by the simultaneous presence of chromosomes 7D and $7H^{ch}$ from *Hordeum chilense*. *Crop. Pasture Sci.* **2015**, *66*, 912–921. [CrossRef]
50. Friebe, B.; Jiang, J.; Raupp, W.J.; McIntosch, R.A.; Gill, B.S. Characterization of wheat alien translocations conferring resistance to diseases and pests: Current status. *Euphytica* **2006**, *91*, 59–87. [CrossRef]
51. Qi, L.; Friebe, B.; Zhang, P.; Gill, B.S. Homoeologous recombination, chromosome engineering and crop improvement. *Chromosome Res.* **2007**, *15*, 3–19. [CrossRef]

Article

Wx Gene in *Hordeum chilense:* Chromosomal Location and Characterisation of the Allelic Variation in the Two Main Ecotypes of the Species

Juan B. Alvarez *, Laura Castellano, Rocío Recio and Adoración Cabrera

Departamento de Genética, Escuela Técnica Superior de Ingeniería Agronómica y de Montes, Edificio Gregor Mendel, Campus de Rabanales, Universidad de Córdoba, CeiA3, ES-14071 Córdoba, Spain; a02caall@uco.es (L.C.); rorecas@gmail.com (R.R.); ge1cabca@uco.es (A.C.)

* Correspondence: jb.alvarez@uco.es

Received: 17 April 2019; Accepted: 17 May 2019; Published: 22 May 2019

Abstract: Starch, as the main grain component, has great importance in wheat quality, with the ratio between the two formed polymers, amylose and amylopectin, determining the starch properties. Granule-bound starch synthase I (GBSSI), or waxy protein, encoded by the *Wx* gene is the sole enzyme responsible for amylose synthesis. The current study evaluated the variability in *Wx* genes in two representative lines of *Hordeum chilense* Roem. et Schult., a wild barley species that was used in the development of tritordeum (×*Tritordeum* Ascherson et Graebner). Two novel alleles, *Wx-H*^{ch}*1a* and *Wx-H*^{ch}*1b*, were detected in this material. Molecular characterizations of these alleles revealed that the gene is more similar to the *Wx* gene of barley than that of wheat, which was confirmed by phylogenetic studies. However, the enzymatic function should be similar in all species, and, consequently, the variation present in *H. chilense* could be utilized in wheat breeding by using tritordeum as a bridge species.

Keywords: starch; tritordeum; waxy proteins; wheat quality; wild barley

1. Introduction

Starch is the main component of wheat grain, constituting up to 75% of its dry weight. This polysaccharide contains two different glucose polymers: amylose (22%–35% of the total) and amylopectin (68%–75% of the total) [1]. Changes in the ratio between these polymers have a clear influence on starch gelatinization, pasting and gelation properties [2], affecting the end-use quality levels of different wheat products, such as bread, pasta, and noodles [3–5], as well their shelf-lives [6] and nutritional values [7].

Starch synthesis involves several starch synthases, starch branching enzymes, and starch debranching enzymes [8]. The most studied of these proteins has been the granule-bound starch synthase I (GBSSI) or waxy proteins (ADP glucose starch glycosyl transferase, EC 2.4.1.21), which are solely responsible for amylose synthesis [9]. In wheat, these proteins are synthesized by genes located in the short arm of the seven-group homeologous chromosome, with the exception of the *Wx-B*1 gene that, owing to a translocation event, is located in the 4AL chromosome [10]. In wheat relatives, this gene is located in similar positions, and its molecular configuration of 12 exons and 11 introns is highly conserved in all of these species [11]. In other Poaceae species, such as barley (*Hordeum vulgare* L.), this gene has shown the identical structure and location [12].

The variability of waxy proteins has been studied in common and durum wheat, as well as in some wild and cultivated relatives [13]. However, the variability in modern wheat cultivars is not very wide, according to data in the Wheat Gene Catalogue [14]. In the search for new waxy variants, species from the primary and secondary wheat pools could contain good candidates. These species have been successfully used to transfer useful traits to wheat. In some cases, these transfer events have generated amphiploids that have also been used as bridge species [15]. In other cases, these amphiploids have been derived to produce human-made crops, such as tritordeum (×*Tritordeum* Ascherson et Graebner), that have shown promising characteristics [16].

Tritordeum was synthesized using mainly durum wheat and *Hordeum chilense* Roem. et Schult. ($2n = 2\times = 14$, $H^{ch}H^{ch}$), a wild barley species native to Chile and Argentina, included in the section *Anisolepis* Nevski [17]. In this species, variation in genes related to quality has been widely evaluated over the last decade [18–21] and has been used to expand the genetic base of tritordeum. Furthermore, this species exhibits advantageous agronomic and quality characteristics [22–24], which, together with its ability to be crossed with other members of the *Triticeae* tribe [25], make it useful in cereal breeding.

In its natural distribution area, *H. chilense* shows some ecotypes, based on morphological and ecophysiological traits [26,27]. The two main groups are related to the first two *H. chilense* lines used to develop tritordeum, H1 and H7. Tritordeum developed using these lines showed differences in fertility, grain size, and life cycle, depending on the female parent (H1 or H7) used [25]. An analysis of the genes related to the flour quality from both lines also showed that these ecotypes could have different effects on the tritordeum quality. These differences have been evaluated for the seed storage proteins [18,28], hordoindolines [29], and pigment enzymes [30,31].

The main goals of this study were to analyze allelic variation and molecularly characterize of the *Wx* genes in the H1 and H7 lines of *H. chilense*, and to determine the gene's chromosomal location.

2. Materials and Methods

2.1. Plant Materials

Seeds of two *H. chilense* lines (H1 and H7) that were self-pollinated for two generations were used in this study. The ditelosomic addition lines (CS + $7H^{ch}S$ and CS + $7H^{ch}L$), together with both parental lines (common wheat cv. "Chinese Spring" and line H1) were used to locate the *Wx* gene in *H. chilense*. These materials were grown in greenhouse conditions.

2.2. DNA Extraction and PCR Amplification

For DNA extractions, ~100 mg of young leaf tissue was excised and immediately frozen in liquid nitrogen. DNA was isolated using the cetyltrimethyl ammonium bromide (CTAB) method as described by Stacey and Isaac [32].

The primers BDFL (5′-CTGGCCTGCTACCTCAAGAGCAACT-3′) and BRD (5′-CTGACGTCCATGCCGTTGACGA-3′) designed by Nakamura et al. [33] were used to detect the presence of the *Wx-H^{ch}1* gene in the ditelosomic addition lines. The amplification was performed in a 20 μL final reaction volume, containing 50 ng of genomic DNA, 1.25 mM $MgCl_2$, 0.2 mM dNTPs, 4 μL 10× PCR buffer, 0.2 μM of each primer and 0.75 U GoTaq® G2 Flexi DNA polymerase (Promega). The PCR conditions included an initial denaturation step of 3 min at 94 °C followed by 35 cycles as follows: 30 s at 94 °C, 30 s at 65 °C then 2 min at 72 °C. After the 35 cycles, a final extension of 5 min at 72 °C was included.

Amplification products were fractionated in vertical PAGE gels with 8% polyacrylamide concentration (*w*/*v*, C: 1.28%) and the bands were stained with GelRed™ nucleic acid staining (Biotium) and visualized under UV light.

2.3. Cloning of PCR Products and Sequencing Analysis

Owing to the length and structure of the *Wx* gene, ~2800 bp with 11 introns and 12 exons, three fragments were amplified using primers designed by Guzmán and Alvarez [34]. The first fragment includes the first to third exons (Wx1Fw: 5′-TTGCTGCAGGTAGCCACACC-3′ and Wx1Rv: 5′-CCGCGCTTGTAGCAGTGGAA-3′), the second extends from the third to the sixth exon (Wx2Fw: 5′-ATGGTCATCTCCCCGCGCTA-3′ and Wx2Rv: 5′-GTTGACGGCGAGGAACTTGT-3′), and the last fragment covers the region spanning the 6th to the 11th exon (Wx3Fw: 5′-GGCATCGTCAACGGCATGGA-3′ and Wx3Rv: 5′-TTCTCTCTTCAGGGAGCGGC-3′).

All amplifications were performed in 50 µL final volumes, containing 100 ng of DNA genomic, 1.25 mM $MgCl_2$, 0.2 mM dNTPs, 10 µL 10× PCR buffer and 0.75 U GoTaq®G2 Flexi DNA polymerase (Promega). The primer concentrations were 0.4, 0.3 and 0.2 µM per primer for the first, second and third fragments, respectively. The PCR conditions included an initial denaturation step of 3 min at 94 °C and then 35 cycles as follows: for Wx1Fw/Wx1Rv, 40 s at 94 °C, 30 s at 64 °C and 1 min at 72 °C, for Wx2Fw/Wx2Rv, 30 s at 94 °C, 30 s at 66 °C and 90 s at 72 °C, and for *Wx3Fw/Wx3Rv*, 40 s at 94 °C, 30 s at 62 °C and 90s at 72 °C. After the 35 cycles, all reactions included a final extension of 5 min at 72 °C.

The PCR products were purified by separation in 1% agarose gel, excised and then independently ligated into the pSpark®-TA Done vector (Canvax). They were then transformed into *Escherichia coli* 'CVX5α' competent cells (Canvax). Inserts were sequenced by Sanger method from at least three different clones. The novel sequences are available from the GenBank database [*Wx-H^{ch}1a*: MK045501 for the H1 line, and *Wx-H^{ch}1b*: MK045502 for the H7 line].

2.4. Data Analysis

The sequences were analyzed and compared with sequences of CS (*Wx-A*1: AB019622, *Wx-B1*: AB019623, and *Wx-D1*: AB019624), two-rowed barley cv. Vogelsanger Gold (*Wx-H*1: X07931) and six-rowed barley cv. Morex (*Wx-H1*: AF474373) available in the databases using Geneious Pro version 5.0.4 software (Biomatters Ltd., Auckland, New Zealand). The synonymous substitution rate (*Ks*) and non-synonymous substitution rate (*Ka*), as well as the *Ka/Ks* ratios, were computed using DNAsp ver. 5.0 [35]. Divergence times were calculated by the mean divergence time, 2.7 million years ago (MYA) between the A and D genomes estimated by Dvorak and Akhunov [36]. Predicted proteins, as well as secondary structure predictions, were also obtained with this software, using the EMBOSS tool Garnier [37]. Amino acid substitutions between the predicted proteins obtained from the new alleles and reference proteins were analyzed using the PROVEAN (Protein Variation Effect Analyzer) software tool to predict whether these amino acid substitutions or InDels have an impact on their biological function [38,39].

A phylogenetic tree was constructed with the software MEGA6 [40] using the complete coding regions of the two sequences obtained, together with the following sequences of the *Wx* genes: common wheat CS (*Wx-A*1: AB019622, *Wx-B*1: AB019623, and *Wx-D*1: AB019624), two-rowed barley "Vogelsanger Gold" (*Wx-H*1: X07931) and six-rowed barley 'Morex' (*Wx-H*1: AF474373). A neighbor-joining cluster of all of the analyzed sequences was generated using the Poisson correction method for amino acid sequences [41] with one bootstrap consensus from 1,000 replicates [42].

3. Results

To determine the location of the *Wx* gene, different combinations of primers were used. However, for some of them, owing to similar fragment sizes, establishing the unambiguous presence of the *Wx* gene from *H. chilense* was difficult. The BDFL/BRD primers, designed by Nakamura et al. [33], provided the most reliable results (Figure 1).

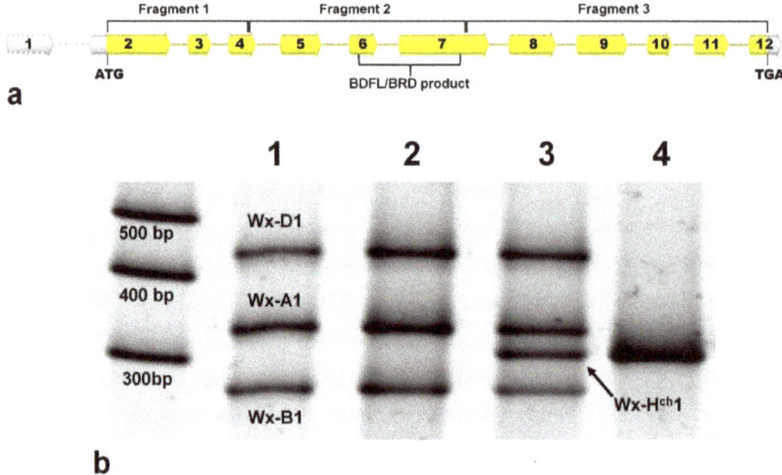

Figure 1. (**a**) Diagrammatic representation of *Wx* gene showing the three fragments used for sequencing, and (**b**) PCR analysis for the chromosomal location of *H. chilense Wx* gene using primers BDFL/BRD from Nakamura et al. [33] in common wheat, ditelosomic addition lines, and *H. chilense.* Lanes are as follows: 1, cv. Chinese Spring (CS), 2, CS + $7H^{ch}S$ line, 3, CS + $7H^{ch}L$ line, and 4, *H1* line.

Because data obtained from other *Triticeae* species indicated that the *Wx* genes were mainly located on the short arm of the chromosome 7 [10], the CS + $7H^{ch}S$ and CS + $7H^{ch}L$ lines were used to determine the arm location of this gene in *H. chilense*. Figure 1 shows the presence of one additional band in the CS + $7H^{ch}L$ line (Lane 3) that is a similar size to a band in the *H. chilense* (H1) line (Lane 4), which is absent in both the common wheat CS (Lane 1) and the CS + $7H^{ch}S$ line (Lane 2). Thus, the *Wx-H^ch^1* gene is located on $7H^{ch}L$ (H1), suggesting an inversion in this *H. chilense* chromosome.

The *Wx-*H^{ch}*1* gene was analyzed in two lines of *H. chilense* that represent two different biotypes of this species present in Chile. Due to the length of this gene (~2700 bp), the genomic sequence was obtained amplifying three fragments, which covered the complete coding sequence. The first fragment of ~620 bp, covered part of the second exon (the first one in the coding sequence, see Figure 1) until the end of the fourth exon, while the second fragment (~960 bp) spanned the fourth to the seventh exons. Finally, the third fragment (~1160 bp) was the region between the end of fragment 2 and the 12th exon, including the TGA codon. The alignment and comparison are shown in Figure S1. The initiation codon, ATG, and the termination codon, TGA, for translation, as well as the splice junctions of each intron of *Wx-*H^{ch}*1*, were in homologous positions to those in other *Wx* genes. Both alleles detected in *H. chilense* (*Wx-*H^{ch}*1a* and *Wx-*H^{ch}*1b*) were smaller in size than the *Wx* genes used for comparison (Table 1).

The comparison between the seven nucleotide sequences showed that the greatest homology level was detected between the *Wx-*H^{ch}*1* genes and the *Wx-H1* variants from barley (90.8%). However, the comparison of the *Wx* genes from common wheat cv. Chinese Spring showed lower values of 83.6% for *Wx-D1*, 84.7% for *Wx-B1* and 87.7% for *Wx-A1*. Nevertheless, the predicted proteins of these same sequences showed homology greater than 94% for all comparisons, and greater than 97.7% among barley species. This is in concordance with most of the sequence differences being found in introns. In all cases, the exons were the same size, with the exception of exon 2, which contained one or two additional codons in wheat but was similar in both *H. chilense* and *H. vulgare* (Table 1).

Table 1. Size of the different exons and introns of the coding sequence in the *Wx* sequences evaluated.

	Wx-A1a [1]	*Wx-B1a* [1]	*Wx-D1a* [1]	*Wx-H1a* [2]/*Wx-H1b* [3]	*Wx-H^ch1a*/*Wx-H^ch1b*
Exon 2	321	324	321	318	318
Exon 3	81	81	81	81	81
Exon 4	99	99	99	99	99
Exon 5	154	154	154	154	154
Exon 6	101	101	101	101	101
Exon 7	354	354	354	354	354
Exon 8	180	180	180	180	180
Exon 9	192	192	192	192	192
Exon 10	87	87	87	87	87
Exon 11	129	129	129	129	129
Exon 12	117	117	117	117	117
Intron 2	82	99	90	89	85
Intron 3	84	88	95	84	80
Intron 4	109	113	104	126	109
Intron 5	125	133	152	136	113
Intron 6	99	69	141	106	89
Intron 7	91	92	85	92	89
Intron 8	95	86	82	94	94
Intron 9	90	84	84	82	82
Intron 10	98	97	98	97	97
Intron 11	93	115	116	76	85/86
Total	2781	2794	2862	2794	2735/2736

[1] cv. Chinese Spring (NCBI ID: *Wx-A1*, AB019622, *Wx-B1*, AB019623, *Wx-D1*, AB019624) [43]. [2] cv. Vogelsanger Gold (NCBI ID: X07931) [12]. [3] cv. Morex (NCBI ID: AF474373).

3.1. Amino acid Predicted Sequence Analysis

While coding sequences of the *Wx* genes varied, most variation resulted in silent mutations that did not impact protein sequence or structure. Additionally, these proteins were synthesized as precursors or pre-proteins, including one transit-peptide of 70 amino acids and one mature domain. Nevertheless, potentially impactful sequence variation was detected in a conserved region of the mature domain related to waxy protein activity. These changes can lead to marked differences in the predicted sequences of the respective proteins (Figure S2).

The *Hordeum* sequences, including both *H. chilense* biotypes, showed the insertion of one amino acid residue within the signal peptide between positions 53 and 54. Furthermore, these sequences had deletions of Gly73 or Ala73 residues detected in the wheat proteins. Both InDels were the consequences of the aforementioned elimination of one or two codons in exon 2. Three non-conservative amino acid changes were detected in both *H. chilense* variants. For two of these changes, Pro24 → Arg and Ser416 → Pro, the H7 line showed the same amino acids as the other evaluated sequences. The Ser416 → Pro change could have deleterious effects according to the PROVEAN analysis, with a score of −2.869. The 419 position was different in both *H. chilense* sequences and was also different than the other sequences, with the exception of Wx-A1, which was similar to that of Wx-H^{ch}1a (Table 2).

The amino acid sequences from *H. chilense* were more similar to the waxy proteins from barley than those derived from any wheat genome. Only 15 changes were detected between *H. chilense* and barley variants, while up to 54 changes were observed when this comparison was carried out with wheat waxy proteins, and ten changes were common to both barley and wheat species (Figure S2).

Up to 15 amino acids variants were detected within the transit peptide. The sequence changes generated some variations in the secondary structure of the transit peptide. The most dramatic was the Val5 → Ala change (detected in barley waxy proteins but not in *H. chilense* ones) that resulted in an elongated first helix and the disappearance of a β strand in the secondary structure (Figure 2). With the exception of the aforementioned change in position 24 (Pro in Wx-H^{ch}1a and Arg in the others), all changes were common to both *H. chilense* variants. The barley variants showed three of these conservative changes (Val5 → Ala, Ile18 → Val and Met68 → Val), although the waxy protein of cv. Vogelsanger Gold (*Wx-H1a* variant) included one additional non-conservative change at position 70 (Arg70 → Ser). Four of these differences, all classified as conservative, were common to Wx-B1 and Wx-D1 (Pro25 → Ala, Leu30 → Val, Asn34 → Ser, and Ser62 → Thr), although each variant showed two additional changes (Ala39 → Pro and Ile45 → Thr for Wx-B1, and Ile45 → Val and Lys52 → Thr for Wx-D1). Nevertheless, Wx-A1 was the most varied, with five unique changes (three conservative: Ile18 → Val, Ala58 → Pro and Gly61 → Phe, and two non-conservative: Gly17 → Ser and Ser62 → Asp) and one change in common with Wx-D1 (Ile45 → Val) (Figure 2).

Table 2. Amino acid comparison of predicted mature protein among waxy protein variants evaluated.

Position [1]	*Wx-H^ch1a/b*	*Wx-H1a* [2]	*Wx-A1a* [2]	*Wx-B1a* [2]	*Wx-D1a* [2]
103	Pro		Ala		
115	Val		Ile	Ile	Ile
123	Asn	Lys	Lys	Lys	Lys
131	Val	Ile	Ile		
137	Ala		Val		Val
139	Glu		Arg		Lys
142	Thr	Arg	Arg	Arg	Arg
145	Phe		Tyr	Tyr	Tyr
158	Ile		Val	Val	Val
162	Trp		Cys	Cys	Cys
189	Gln			Leu	
201	Ala		Val		Val
206	Asp	Asn			Asn
208	Asn				Asp
212	Tyr		His		
232	Pro	Leu	Leu	Leu	Leu
244	Asn			Ser	
249	Thr				Ala
356	Thr		Ile	Ala	
362	Ala		Thr		
363	Val			Ala	
367	Ile		Val	Val	
373	Ala		Gly	Gly	Gly
416	Ser/Pro	Pro	Pro	Pro	Pro
419	Val/Met	Leu		Leu	Leu
427	Ile		Val	Val	Val
438	Arg	Lys			
443	Val	Met			Ile
449	Gly		Thr	Ser	Ser
452	Arg		Trp		
471	Leu		Val	Val	Val
496	Ala	Val			
508	Val			Met	
535	Ala		Val	Val	Val
551	Gln		His	His	His
587	Ile			Val	Val
588	Val			Ile	Ile
590	Asp	Glu	Glu	Glu	Glu
597	Met		Leu		

[1] This position should be increased for *Wx-A1* and -D1 (+1), and for *Wx-B1* (+2). [2] NCBI ID: barley [X07931] and common wheat cv. "Chinese Spring" [*Wx-A1*: AB019622, *Wx-B1*: AB019623, *Wx-D1*: AB019624].

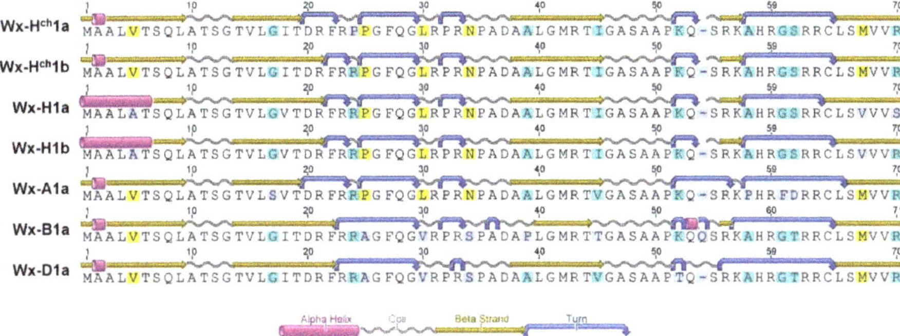

Figure 2. Comparison of the secondary structure motifs predicted by Garnier for the transit peptide region among all sequences evaluated.

Numerous changes were also observed in the mature proteins both in barley and wheat (Table 2). Five of these changes were common to both species (Asn123 → Lys, Thr142 → Arg, Pro 232 → Leu, Ser416 → Pro and Asp590 → Glu), whereas two changes were exclusively detected in barley (Arg438 → Lys and Ala496 → Val) and nine were exclusively detected in wheat (Val115 → Ile, Phe145 → Tyr, Ile158 → Val, Trp162 → Cys, Ala373 → Gly, Ile427 → Val, Leu471 → Val, Ala535 → Val and Gln551 → His). The other changes were detected in one or two wheat sequences (Table 2). Two of these changes could have effects on enzyme function (Figure 3). In addition to the abovementioned change, Ser416 → Pro, which was unique to Wx-H^{ch}1a, another change with deleterious effects predicted by the PROVEAN analysis was Pro232 → Leu, with a score of −4.061. The other changes observed were considered neutral.

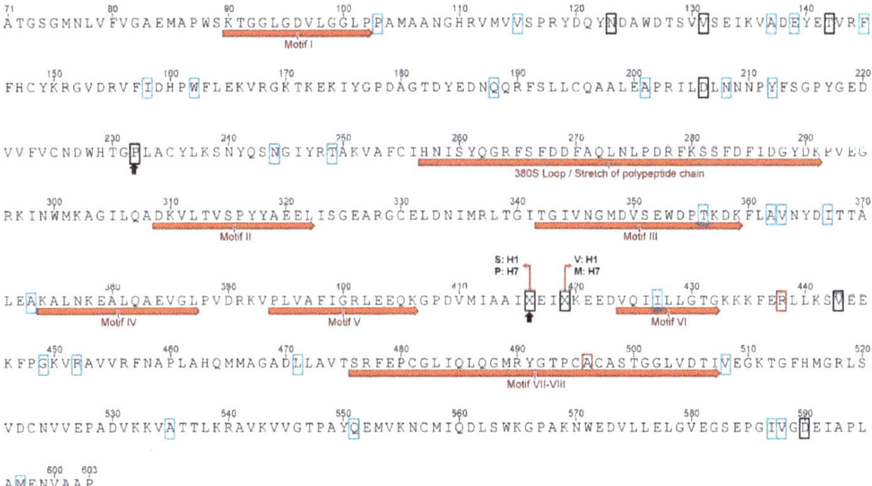

Figure 3. Consensus sequence of the predicted proteins from *H. chilense* showing the motifs described by Leterrier et al. [44] conserved in waxy proteins. Squares indicate substitution sites (blue for wheat, red for barley and black for both ones), and arrows point relevant changes found in the novel alleles.

Up to five of the eight motifs described by Leterrier et al. [44] are involved in the ADP glucose-binding and catalytic sites within the mature protein. Three changes were observed inside these conserved motifs (Figure 3). However, only the change Ile427 → Val was considered relevant because, although this change was also detected in barley waxy protein, the wheat waxy proteins all showed Val as the residue in this position. The change Thr356 appeared in barley and the *Wx-D1* protein, while Val 496 was exclusively found in the barley protein. The PROVEAN analysis suggested that these changes be considered neutral because their influence on the enzyme function was very limited.

3.2. Phylogenetic Analysis

The complete amino acid sequences of the two novel variants from *H. chilense* obtained in this study, together with other waxy protein sequences present in the NCBI database, were used to construct a phenogram based on the Poisson correction method for amino acid sequences (Figure 4). Three main clusters were observed, representing the correlations between the *H. chilense* sequences and the common barley sequences used.

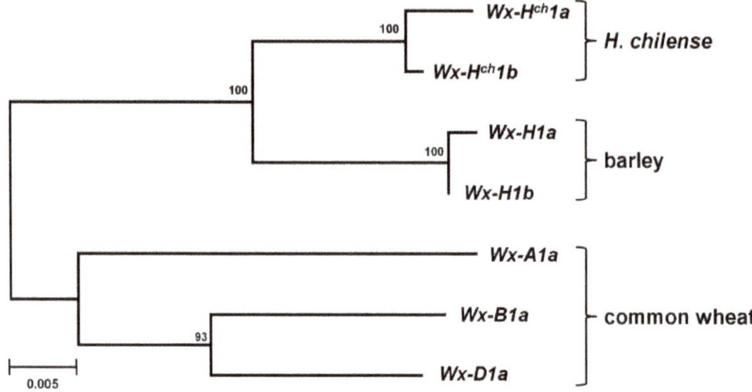

Figure 4. Neighbor-joining tree based on the Poisson correction method for amino acid sequences analyzed. Number above nodes indicates bootstrap estimates from 1000 replications.

These data were corroborated when the genomic nucleotide sequences of these variants were analyzed. Furthermore, the *Ks* and *Ka* substitution rates among *Wx* genes were calculated using the coding sequences of the complete genes. The comparison value between the genes from *H. chilense* and *H. vulgare* was high (*Ks* = 0.127), which suggested that the divergence time between species was ~2.3 MYA based on the mean divergence rate (0.0533 *Ks* per MY) obtained for this gene in a previous study [45].

4. Discussion

Knowledge regarding the influence of the amylose/amylopectin ratio on starch properties has encouraged the search for allelic variants that could increase/decrease either starch component. The most studied, in this context, has been the ADP glucose starch glycosyl transferase (GBSSI or the waxy proteins) solely responsible for amylose synthesis. This starch synthase has been studied in several cereal species, mainly those in which the starch properties are important for their use in the agri-food industry or in bio-ethanol production [46].

Recently, biotechnological techniques have allowed the development of new species using phylogenetically related species. These new species could also be used as a bridge to transfer new variations to common wheat. Thus, *H. chilense*, as a species involved in the synthesis of tritordeum, could be useful [25]. The incorporation of the H^{ch} genome in durum wheat has clear effects on the quality characteristics of the tritordeum. For example, the presence of the glutenins or hordeins of this wild species modifies the strength of the gluten in tritordeum flour [22], and their hordoindolines change the texture of the grain from the ultra-hard of the parent durum wheat to soft in the derived tritordeum [29].

Here, we have studied one of the main keys in cereal flour quality, starch, by molecularly characterising the *Wx* gene in the two main lines of *H. chilense* used in the development of the tritordeums [25]. The *H. chilense* waxy proteins presented structures very similar to those of waxy proteins in other species of *Triticeae*, such as wheat and barley. The sizes of the predicted proteins were similar, although numerous amino acid changes were detected. However, these changes were mostly silent and not related to the active site of this enzyme, and probably, without influence on its function. In fact, some of these changes have been observed in other *Wx* genes [13]. The highly conserved structure of this gene makes it a good candidate for phylogenetic analysis [11,45,47–51]. In this study, the use of *Ka* established the separation between the *Hordeum* genomes at ~3 MYA.

In barley, Kramer and Blander [52] located the *Wx* gene on the short arm of chromosome 1 (7H). In common wheat, the waxy loci are located on chromosome 7AS (*Wx*-A1), chromosome 4AL, which was translocated from the original 7BS, (*Wx*-B1) and chromosome 7DS (*Wx*-D1) [10]. Here, the $Wx\text{-}H^{ch}1$ gene from *H. chilense* was located on $7H^{ch}L$, opposite the arm location found in the other Triticeae species [13]. Mattera et al. [53] indicated a similar change in the location of the *Phytoene syntase* (*Psy-1*) gene in *H. chilense. Psy*-1 was mapped in the distal region of $7\text{H}^{\text{ch}}\text{S}$, while this gene was located on the opposite arm of chromosome 7 in other Poaceae species [14]. On the basis of these changes, Mattera et al. [53] suggested that an inversion occurred between the distal parts of $7H^{ch}S$ and $7H^{ch}L$, which has been confirmed by Avila et al. [54]. The location of the $Wx\text{-}H^{ch}1$ gene on $7H^{ch}L$ in the present study supports this hypothesis on a structural change involving the distal regions of *H. chilense* chromosome $7H^{ch}$.

5. Conclusions

Variability in the *Wx* gene sequences was detected in two *H. chilense* lines representative of the two main ecotypes of species. This gene is located on the long arm of the $7H^{ch}$ chromosome, opposite to the other *Triticeae* species, which suggests the presence of an inversion between the distal parts of $7H^{ch}S$ and $7H^{ch}L$. Molecular characterization of these alleles showed that this gene is more similar to the *Wx* gene of barley than those of wheat. However, the enzymatic function would be similar in all species and, consequently, the variation present in *H. chilense* could be utilized in wheat breeding through the use of tritordeum as a bridge species.

Supplementary Materials: The following are available online at http://www.mdpi.com/2073-4395/9/5/261/s1, Figure S1. Alignment of nucleotide sequences of the *Wx* alleles evaluated in this study, Figure S2. Alignment of predicted protein sequences of the waxy proteins evaluated in this study.

Author Contributions: J.B.A. conceived and designed the study; J.B.A., L.C., and R.R. performed the experiments; J.B.A. and A.C. analyzed the data and wrote the paper. All authors have read and approved the final manuscript.

Funding: This research was supported by grants AGL2014-52445-R and RTI2018-093367-B-I00 from the Spanish State Research Agency (Ministry of Science, Innovation and Universities), co-financed by the European Regional Development Fund (FEDER) from the European Union.

Conflicts of Interest: The authors declare no conflict of interest.

References

1. James, M.G.; Denyer, K.; Myers, A.M. Starch synthesis in the cereal endosperm. *Curr. Opin. Plant Biol.* **2003**, *6*, 215–222. [CrossRef]
2. Zeng, M.; Morris, C.F.; Batey, I.L.; Wrigley, C.W. Sources of variation for starch gelatinization, pasting, and gelation properties in wheat. *Cereal Chem.* **1997**, *74*, 63–71. [CrossRef]
3. Martin, J.M.; Sherman, J.D.; Lanning, S.P.; Talbert, L.E.; Giroux, M.J. Effect of variation in amylose content and puroindoline composition on bread quality in a hard spring wheat population. *Cereal Chem.* **2008**, *85*, 266–269. [CrossRef]

4. Miura, H.; Tanii, S. Endosperm starch properties in several wheat cultivars preferred for Japanese noodles. *Euphytica* **1994**, *72*, 171–175. [CrossRef]
5. Park, C.S.; Baik, B.-K. Characteristics of French bread baked from wheat flours of reduced starch amylose content. *Cereal Chem.* **2007**, *84*, 437–442. [CrossRef]
6. Hayakawa, K.; Tanaka, K.; Nakamura, T.; Endo, S.; Hoshino, T. End use quality of waxy wheat flour in various grain-based foods. *Cereal Chem.* **2004**, *81*, 666–672. [CrossRef]
7. Regina, A.; Bird, A.; Topping, D.; Bowden, S.; Freeman, J.; Barsby, T.; Kosar-Hashemi, B.; Li, Z.; Rahman, S.; Morell, M. High-amylose wheat generated by RNA interference improves indices of large-bowel health in rats. *Proc. Natl. Acad. Sci. USA* **2006**, *103*, 3546–3551. [CrossRef] [PubMed]
8. Baldwin, P.M. Starch Granule-Associated Proteins and Polypeptides: A Review. *Starch-Stärke* **2001**, *53*, 475–503. [CrossRef]
9. Echt, C.S.; Schwartz, D. Evidence for the inclusion of controlling elements within the structural gene at the *Waxy* locus in maize. *Genetics* **1981**, *99*, 275–284.
10. Yamamori, M.; Nakamura, T.; Endo, T.R.; Nagamine, T. Waxy protein deficiency and chromosomal location of coding genes in common wheat. *Theor. Appl. Genet.* **1994**, *89*, 179–184. [CrossRef]
11. Mason-Gamer, R.J.; Weil, C.F.; Kellogg, E.A. Granule-bound starch synthase: Structure, function, and phylogenetic utility. *Mol. Biol. Evol.* **1998**, *15*, 1658–1673. [CrossRef] [PubMed]
12. Rohde, W.; Becker, D.; Salamini, F. Structural analysis of the waxy locus from *Hordeum vulgare*. *Nucleic Acids Res.* **1988**, *16*, 7185–7186. [CrossRef]
13. Guzmán, C.; Alvarez, J.B. Wheat waxy proteins: Polymorphism, molecular characterization and effects on starch properties. *Theor. Appl. Genet.* **2016**, *129*, 1–16. [CrossRef] [PubMed]
14. McIntosh, R.A.; Yamazaki, Y.; Dubcovsky, J.; Rogers, W.J.; Morris, G.; Appels, R.; Xia, X.C. Catalogue of gene symbols for wheat. 2013. Available online: http://www.shigen.nig.ac.jp/wheat/komugi/genes/macgene/2013/GeneSymbol.pdf. (accessed on 23 August 2018).
15. Alvarez, J.B.; Guzmán, C. Interspecific and intergeneric hybridization as a source of variation for wheat grain quality improvement. *Theor. Appl. Genet.* **2018**, *131*, 225–251. [CrossRef] [PubMed]
16. Martín, A.; Alvarez, J.B.; Martín, L.M.; Barro, F.; Ballesteros, J. The development of tritordeum: A novel cereal for food processing. *J. Cereal Sci.* **1999**, *30*, 85–95. [CrossRef]
17. Von Bothmer, R.; Jacobsen, N.; Baden, C.; Jorgensen, R.; Linde-Laursen, I. *An Ecogeografical Study of the Genus Hordeum*, 2nd ed.; International Plant Genetic Resources Institute: Rome, Italy, 1995.
18. Alvarez, J.B.; Martín, A.; Martín, L.M. Variation in the high-molecular-weight glutenin subunits coded at the *Glu-H^{ch}1* locus in *Hordeum chilense*. *Theor. Appl. Genet.* **2001**, *102*, 134–137. [CrossRef]
19. Alvarez, J.B.; Broccoli, A.; Martín, L.M. Variability and genetic diversity for gliadins in natural populations of *Hordeum chilense* Roem. et Schult. *Genet. Resour. Crop Evol.* **2006**, *53*, 1419–1425. [CrossRef]
20. Atienza, S.G.; Giménez, M.J.; Martín, A.; Martín, L.M. Variability in monomeric prolamins in *Hordeum chilense*. *Theor. Appl. Genet.* **2000**, *101*, 970–976. [CrossRef]
21. Atienza, S.G.; Alvarez, J.B.; Villegas, A.M.; Gimenez, M.J.; Ramirez, M.C.; Martín, A.; Martín, L.M. Variation for the low-molecular-weight glutenin subunits in a collection of *Hordeum chilense*. *Euphytica* **2002**, *128*, 269–277. [CrossRef]
22. Alvarez, J.B.; Campos, L.A.C.; Martín, A.; Martín, L.M. Influence of HMW and LMW glutenin subunits on gluten strength in hexaploid tritordeum. *Plant Breed.* **1999**, *118*, 456–458. [CrossRef]
23. Martinek, P.; Svobodová, I.; Věchet, L. Selection of the wheat genotypes and related species with resistance to *Mycosphaerella graminicola*. *Agriculture* **2013**, *59*, 65–73. [CrossRef]
24. Rodríguez-Suárez, C.; Mellado-Ortega, E.; Hornero-Méndez, D.; Atienza, S. Increase in transcript accumulation of *Psy1* and *e-Lcy* genes in grain development is associated with differences in seed carotenoid content between durum wheat and tritordeum. *Plant Mol. Biol.* **2014**, *84*, 659–673. [CrossRef] [PubMed]
25. Martín, A.; Martín, L.M.; Cabrera, A.; Ramírez, M.C.; Giménez, M.J.; Rubiales, D.; Hernández, P.; Ballesteros, J. The potential of *Hordeum chilense* in breeding Triticeae species. In *Triticeae III*; Jaradat, A.A., Humphreys, M., Eds.; Science Publishers Inc.: Enfield, UK, 1998; pp. 377–386.
26. Tobes, N.; Ballesteros, J.; Martínez, C.; Lovazzano, G.; Contreras, D.; Cosio, F.; Gastó, J.; Martín, L.M. Collection mission of *H. chilense* Roem. et Schult. in Chile and Argentina. *Genet. Resour. Crop Evol.* **1995**, *42*, 211–216. [CrossRef]

27. Giménez, M.J.; Cosío, F.; Martínez, C.; Silva, F.; Zuleta, A.; Martín, L.M. Collecting *Hordeum chilense* Roem. et Schult. germplasm in desert and steppe dominions of Chile. *Plant Genet. Resour. Newsl.* **1997**, *109*, 17–19.
28. Pistón, F.; Shewry, P.; Barro, F. D hordeins of *Hordeum chilense*: A novel source of variation for improvement of wheat. *Theor. Appl. Genet.* **2007**, *115*, 77–86. [CrossRef]
29. Guzmán, C.; Alvarez, J.B. Molecular characterization of two novel alleles of *Hordoindoline* genes in *Hordeum chilense* Roem. et Schult. *Genet. Resour. Crop Evol.* **2014**, *61*, 307–312. [CrossRef]
30. Alvarez, J.B.; Martín, L.M.; Martín, A. Chromosomal localization of genes for carotenoid pigments using addition lines of *Hordeum chilense* in wheat. *Plant Breed.* **1998**, *117*, 287–289. [CrossRef]
31. Rodríguez-Suárez, C.; Atienza, S.G.; Pistón, F. Allelic variation, alternative splicing and expression analysis of *Psy1* gene in *Hordeum chilense* Roem. et Schult. *PLoS ONE* **2011**, *6*, e19885. [CrossRef] [PubMed]
32. Stacey, J.; Isaac, P.G. Isolation of DNA from Plants. In *Protocols for Nucleic Acid Analysis by Nonradioactive Probes*; Isaac, P.G., Ed.; Humana Press: Totowa, NJ, USA, 1994; pp. 9–15.
33. Nakamura, T.; Vrinten, P.; Saito, M.; Konda, M. Rapid classification of partial waxy wheats using PCR-based markers. *Genome* **2002**, *45*, 1150–1156. [CrossRef]
34. Guzman, C.; Alvarez, J.B. Molecular characterization of a novel waxy allele (*Wx-A^{u}1a*) from *Triticum urartu* Thum. ex Gandil. *Genet. Resour. Crop Evol.* **2012**, *59*, 971–979. [CrossRef]
35. Librado, P.; Rozas, J. DnaSP v5: A software for comprehensive analysis of DNA polymorphism data. *Bioinformatics* **2009**, *25*, 1451–1452. [CrossRef] [PubMed]
36. Dvorak, J.; Akhunov, E.D. Tempos of Gene Locus Deletions and Duplications and Their Relationship to Recombination Rate During Diploid and Polyploid Evolution in the Aegilops-Triticum Alliance. *Genetics* **2005**, *171*, 323–332. [CrossRef] [PubMed]
37. Rice, P.; Longden, I.; Bleasby, A. EMBOSS: The European molecular biology open software suite. *Trends Genet.* **2000**, *16*, 276–277. [CrossRef]
38. Choi, Y.; Chan, A.P. PROVEAN web server: A tool to predict the functional effect of amino acid substitutions and indels. *Bioinformatics* **2015**, *31*, 2745–2747. [CrossRef] [PubMed]
39. Choi, Y.; Sims, G.E.; Murphy, S.; Miller, J.R.; Chan, A.P. Predicting the Functional Effect of Amino Acid Substitutions and Indels. *PLoS ONE* **2012**, *7*, e46688. [CrossRef]
40. Tamura, K.; Stecher, G.; Peterson, D.; Filipski, A.; Kumar, S. MEGA6: Molecular Evolutionary genetics analysis version 6.0. *Mol. Biol. Evol.* **2013**, *30*, 2725–2729. [CrossRef] [PubMed]
41. Zuckerkandl, E.; Pauling, L. Evolutionary Divergence and Convergence in Proteins. In *Evolving Genes and Proteins*; Bryson, V., Vogel, H.J., Eds.; Academic Press: New York, NY, USA, 1965; pp. 97–166.
42. Felsenstein, J. Confidence limits on phylogenies: An approach using the bootstrap. *Evolution* **1985**, *39*, 783–791. [CrossRef]
43. Murai, J.; Taira, T.; Ohta, D. Isolation and characterization of the three *Waxy* genes encoding the granule-bound starch synthase in hexaploid wheat. *Gene* **1999**, *234*, 71–79. [CrossRef]
44. Leterrier, M.; Holappa, L.; Broglie, K.; Beckles, D. Cloning, characterisation and comparative analysis of a starch synthase IV gene in wheat: Functional and evolutionary implications. *BMC Plant Biol.* **2008**, *8*, 98. [CrossRef]
45. Guzmán, C.; Caballero, L.; Martín, L.M.; Alvarez, J.B. Waxy genes from spelt wheat: New alleles for modern wheat breeding and new phylogenetic inferences about the origin of this species. *Ann. Bot.* **2012**, *110*, 1161–1171. [CrossRef]
46. Cornejo-Ramírez, Y.I.; Martínez-Cruz, O.; Del Toro-Sánchez, C.L.; Wong-Corral, F.J.; Borboa-Flores, J.; Cinco-Moroyoqui, F.J. The structural characteristics of starches and their functional properties. *CyTA-J. Food* **2018**, *16*, 1003–1017.
47. Yan, L.; Bhave, M.; Fairclough, R.; Konik, C.; Rahman, S.; Appels, R. The genes encoding granule-bound starch synthases at the waxy loci of the A, B, and D progenitors of common wheat. *Genome* **2000**, *43*, 264–272. [CrossRef] [PubMed]
48. Mason-Gamer, R.J. Origin of North American *Elymus* (Poaceae: Triticeae) allotetraploids based on Granule-bound starch synthase gene sequences. *Syst. Bot.* **2001**, *26*, 757–768.
49. Ingram, A.L.; Doyle, J.J. The origin and evolution of Eragrostis tef (Poaceae) and related polyploids: Evidence from nuclear waxy and plastid rps16. *Am. J. Bot.* **2003**, *90*, 116–122. [CrossRef] [PubMed]

50. Fortune, P.M.; Schierenbeck, K.A.; Ainouche, A.K.; Jacquemin, J.; Wendel, J.F.; Ainouche, M.L. Evolutionary dynamics of Waxy and the origin of hexaploid Spartina species (Poaceae). *Mol. Phylogenet. Evol.* **2007**, *43*, 1040–1055. [CrossRef] [PubMed]
51. Ortega, R.; Alvarez, J.B.; Guzmán, C. Characterization of the *Wx* gene in diploid *Aegilops* species and its potential use in wheat breeding. *Genet. Resour. Crop Evol.* **2014**, *61*, 369–382. [CrossRef]
52. Kramer, H.H.; Blander, B.A.S. Orienting linkage maps on the chromosomes of barley. *Crop Sci.* **1961**, *1*, 339–342. [CrossRef]
53. Mattera, M.G.; Ávila, C.M.; Atienza, S.G.; Cabrera, A. Cytological and molecular characterization of wheat-*Hordeum chilense* chromosome $7H^{ch}$ introgression lines. *Euphytica* **2015**, *203*, 165–176. [CrossRef]
54. Avila, C.M.; Mattera, M.G.; Rodríguez-Suárez, C.; Palomino, C.; Ramírez, M.C.; Martín, A.; Kilian, A.; Hornero-Méndez, D.; Atienza, S.G. Diversification of seed carotenoid content and profile in wild barley (*Hordeum chilense* Roem. et Schultz.) and *Hordeum vulgare* L.-*H. chilense* synteny as revealed by DArTSeq markers. *Euphytica* **2019**, *215*, 45. [CrossRef]

Article

Analysis of Wheat Bread-Making Gene (*wbm*) Evolution and Occurrence in Triticale Collection Reveal Origin via Interspecific Introgression into Chromosome 7AL

Ilya Kirov [1,*], Andrey Pirsikov [1], Natalia Milyukova [1], Maxim Dudnikov [1], Maxim Kolenkov [1], Ivan Gruzdev [1], Stanislav Siksin [1], Ludmila Khrustaleva [2], Gennady Karlov [1,2] and Alexander Soloviev [1]

1 Laboratory of Marker-Assisted and Genomic Selection of Plants, All-Russia Research Institute of Agricultural Biotechnology, Timiryazevskaya str. 42, 127550 Moscow, Russia; andrey.pyrsikov@yandex.ru (A.P.); milyukovan@gmail.com (N.M.); max.dudnikov.07@gmail.com (M.D.); colenckov@yandex.ru (M.K.); gruzdev82mtz@mail.ru (I.G.); stason_16@inbox.ru (S.S.); karlovg@gmail.com (G.K.); a.soloviev70@gmail.com (A.S.)

2 Center of Molecular Biotechnology, Russian State Agrarian University-Moscow Timiryazev Agricultural Academy, Timiryazevskaya str. 49, 127550 Moscow, Russia; ludmila.khrustaleva19@gmail.com

* Correspondence: kirovez@gmail.com

Received: 11 November 2019; Accepted: 4 December 2019; Published: 5 December 2019

Abstract: Bread-making quality is a crucial trait for wheat and triticale breeding. Several genes significantly influence these characteristics, including glutenin genes and the wheat bread-making (*wbm*) gene. World wheat collection screening showed that only a few percent of cultivars carry the valuable *wbm* variant, providing a useful source for wheat breeding. In contrast, no such analysis has been performed for triticale (wheat (AABB genome) × rye (RR) amphidiploid) collections. Despite the importance of the *wbm* gene, information about its origin and genomic organization is lacking. Here, using modern genomic resources available for wheat and its relatives, as well as PCR screening, we aimed to examine the evolution of the *wbm* gene and its appearance in the triticale genotype collection. Bioinformatics analysis revealed that the wheat Chinese Spring genome does not have the *wbm* gene but instead possesses the orthologous gene, called *wbm-like* located on chromosome 7A. The analysis of upstream and downstream regions revealed the insertion of LINE1 (Long Interspersed Nuclear Elements) retrotransposons and Mutator DNA transposon in close vicinity to *wbm-like*. Comparative analysis of the *wbm-like* region in wheat genotypes and closely related species showed low similarity between the *wbm* locus and other sequences, suggesting that *wbm* originated via introgression from unknown species. PCR markers were developed to distinguish *wbm* and *wbm-like* sequences, and triticale collection was screened resulting in the detection of three genotypes carrying *wbm*-specific introgression, providing a useful source for triticale breeding programs.

Keywords: triticale; wheat bread-making gene; introgression; PCR markers

---

1. Introduction

Wheat is a global source of food and calories and the main ingredient in many products, with an annual world production volume of around 771 million tons as of 2017 (FAOSTAT, 2017). The bread-making quality is the most important trait for wheat breeding programs worldwide. However, the estimation of bread-making quality is challenging because it is costly and requires a substantial amount of seed for analysis [1]. To overcome these difficulties, researchers have searched for the genetic determinants controlling the trait to enable marker-assisted selection (MAS) [2–4]. However, the genetic basis of the bread-making quality of wheat is not yet understood and the genes that have been found do not explain the broad variation in this characteristic. High molecular weight (HMW) glutenin genes (*Glu-A1*, *Glu-B1*, and *Glu-D1*) [2], located on the long arm of group 1 chromosomes, are the most well studied among the major genes controlling bread-making quality. A number of markers have been developed to distinguish different alleles of glutenin genes [5–10]. A gene, called wheat bread-making (*wbm*), was identified in wheat, and its elevated expression level in the endosperm during wheat grain development has been demonstrated [11]. Analysis of *wbm* expression in different wheat genotypes resulted in two contrast groups: (1) genotypes with high expression of *wbm* and (2) genotypes with the negligible expression of the gene. A high correlation of *wbm* expression and the values of some quality characteristics have been demonstrated [1,11]. The expression level of *wbm* along with the allelic composition of glutenin genes, and the presence of 1BL.1RS translocation is one of the key determinants of major wheat quality traits coupled with bread-making quality [1]. Analysis of variation in the upstream sequence of *wbm* allowed the identification of one variant, GWseqVar3, which was associated with high *wbm* expression in cultivars with good bread-making characteristics [11]. Based on this knowledge, PCR-based (NWP, [11]) and Kompetitive allele specific PCR (KASP, [10]) markers were developed to specifically identify GWseqVar3. These markers were employed for screening wheat genotypes in different world collections [1,10]. These studies found rare (0% to 36%) occurrences of GWseqVar3 of *wbm* in wheat collections, highlighting the importance of screening of germplasm on *wbm* to improve bread-making quality.

Despite the relative progress in wheat collection screening, the frequency of *wbm* occurrence in hexaploid triticale (×*Triticosecale* Wittmack) ($2n = 42$, AABBRR, wheat × rye hybrid) collections is unexplored. Triticale occupies 4 million hectares worldwide with an approximate production volume of 15 million ton per year, which is comparable to rye production. Triticale is a hardy crop, combining many beneficial traits inherited from both parents, including high yield, increased tolerance to biotic and abiotic stresses, and nutrient-use efficiency [12–18]. Although triticale is an important crop for biofuel production and forage, application of triticale grains in the bakery industry is limited, with no registered cultivars with good bread-making quality [19]. However, with the current growth in the number of health-conscious people, triticale is becoming more attractive as a human food source [20]. In this context, new breeding strategies are needed to improve triticale end-use characteristics and identification of valuable gene variants located in the A and B genomes.

Study of the *wbm* gene has focused on practical aspects, whereas *wbm* gene origin, evolution, and genomic organization studies are lacking. In their initial publication, Furtado et al. [11] performed PCR screening of wheat progenitor species (*Triticum monococcum*, *Triticum urartu*, and *Triticum turgidum*) to address *wbm* origin but obtained controversial results. The PCR products of the expected size were amplified in some samples of *T. monococcum* (AA genome) and *T. urartu* (AA genome) genomic DNA but not in *T. turgidum* (AABB genome). Because no D-genome donor species (*Aegilops tauschii*) have been used for PCR analysis, the location of the *wbm* gene on wheat A or D genomes was proposed. Rasheed et al. [10], using a Basic Local Alignment Search tool (BLAST) search, proposed the location of the *wbm* gene on chromosome 7AL; whether *wbm* gene can also be found in the D genome has not been studied. Thus, no consistent information about the genomic organization of *wbm* has been reported.

The objective of this study was to examine *wbm* evolution and genomic organization and to use this information in the identification of triticale lines carrying *wbm* for further improvement of the bread-making quality of this crop. From several lines of evidence, we proved the location of *wbm* gene chromosome 7AL and demonstrated considerable diversity in the *wbm* protein sequence between closely related species. Using available genomic data for grasses and PCR analysis with newly developed markers, we found that *wbm* is not an allelic variant but was introgressed from unknown species into the wheat genome. Chinese Spring cultivar has an orthologous *wbm* gene (*wbm-like*), which has negligible expression in grain, and its evolution was accompanied by mobile element insertion. Location of *wbm* in the A genome provides an opportunity for screening the collection of triticale lines possessing A, B, and R genomes, and allowed us to identify three triticale lines possessing this gene.

2. Material and Methods

2.1. Identification of *wbm*-Like Gene

To identify the *wbm* location on a wheat chromosome, a 75-amino acid (aa) sequence of *wbm* protein [11] was employed as a query for tBLASTn search against the expressed sequence tags (EST) collection of NCBI, and an EST clone (BQ606004.1, https://www.ncbi.nlm.nih.gov/nuccore/BQ606004.1/) was found. This EST has a 225-base pair (bp) open reading frame predicted by ORFfinder (https://www.ncbi.nlm.nih.gov/orffinder/), and the corresponding protein sequence was identical to original *wbm* protein. Using this EST sequence, we next conducted a nucleotide BLAST search against the IWGSC RefSeq v1.0 ([21]) reference genome assembly of wheat Chinese Spring (https://urgi.versailles.inra.fr/blast/, [22]) and the NCBI WGS database.

2.2. Multiple Alignment of *wbm* and Related Protein and Nucleotide Sequences

To compare *wbm* and *wbm-like* with the related sequences from other species and cultivars nucleotide BLAST [23] was performed against the latest assembly of their genomes (wheat cultivars: https://wheatis.tgac.ac.uk/grassroots-portal/ [24], http://www.10wheatgenomes.com/; *S. cereale*: https://webblast.ipk-gatersleben.de/ryeselect/, other species: https://plants.ensembl.org/Triticum_aestivum/ [25], https://urgi.versailles.inra.fr/blast/blastresult.php) using default parameters. The *wbm*-similar sequences were retrieved, and open reading frames (ORFs) were predicted using ORFfinder software (https://www.ncbi.nlm.nih.gov/orffinder/). Corresponding ORF sequences overlapping with *wbm* hits and deduced protein sequences were aligned by mafft [26] with standard parameters and visualized using UGene (Version 33, UNIPRO, LLC, Novosibirsk, Russia,) [27].

2.3. Phylogenetic Tree Construction

Phylogenetic trees were build based on multiple alignments of protein or nucleotide sequences of *wbm*, *wbm-like*, and *wbm* similar genes in UGene software [27] using the PhyML maximum likelihood method with bootstrap branch support computed from 100 replicates.

2.4. *wbm*-Like Expression Analysis

The wheat expression atlas (wheat-expression.com [28]) database was mined by the gene corresponding to the *wbm-like* sequence (TraesCS7A02G531903 [21]) in the CS genome. The expression matrix was retrieved only for grain, where *wbm* was found to be expressed [11]. Data were visualized (Supplementary Figure S1) by bar plot build by R programming language using the ggplot2 package [29].

2.5. Plant Material

A total of 107 forms and cultivars of spring triticale were used, including selection forms obtained at the Department of Genetics, Russian State Agrarian University, and accessions from the VIR collection (Supplementary Table S1).

2.6. DNA Isolation and PCR Analysis

PCR amplification was conducted using specific primers (Table 1) designed using Primer 3.0 software (http://www.bioinformatics.nl/cgi-bin/primer3plus/primer3plus.cgi/, [30]). The PCR conditions for custom primers were 94 °C for 1 min, 35 cycles: 94 °C for 1 min; 58 °C for 1 min; 72 °C for 1 min; and final elongation: 72 °C for 3 min. For NWP primers [11], the annealing temperature was 55 °C.

Table 1. Primers designed in this study and the expected length of the PCR product.

Primer ID	Primers 5′-3′	Expected Length of PCR Product (bp)
pro	F: TTGAAGAGAAGTGGCCAGCG R: GTTCATGCGATGCAGAGAGC	988
L1	F: ACAACATCACAAGTGGAAAGGC R: AGGTTGGTAGTAATTCCAAATTCAAGT	2200
wbm	F: TGTGTGTTGCTACCATCATGG R: GGCAGCTCCCATGTTGTACA	219
wbm-like	F: TTACTACGCACGGGCAGTTT R: GTGTGTTGCCACCATCATGG	225

For verification, PCR products of three selected lines were purified and sequenced using ABI 3130×l Genetic Analyzer (Applied Biosystems, Foster City, CA, USA) according to the manufacturer's instructions.

3. Results

3.1. *wbm* Located on Chromosome 7A

We aimed to obtain insight into the genome organization of *wbm* in wheat. For this, we identified the genomic location of *wbm* using corresponding EST and protein sequences [11]. We did not find any genomic sequences with absolute identity with the *wbm* sequence in the wheat Chinese Spring (CS) genome reference data. Hits located on chromosomes 7A and 7D were the most significant (81% identity and 100% coverage, E-value $3e \times 10^{-59}$ for both hits). The hit located on chromosome 7A overlaps a gene, (TraesCS7A02G531903, *wbm-like* hereafter), whereas the hit located on chromosome 7D (*wbm_hit_7D*) partially overlapped with the 5′ end of the annotated non-coding RNA (STRG_Seed.132206.1). We also performed a BLAST search with the Tag-A sequence (CATGTTGTTCCGTGTAGTACC), which was used for *wbm* identification [11]. We found that this sequence had near-perfect similarity (only one mismatch) with the downstream region of *wbm-like*. To obtain additional evidence of *wbm* location on chromosome 7A, we applied the recently released wheat 10+ genomes project (http://www.10wheatgenomes.com/). Using *wbm-like*, *wbm*, and their promoter sequences, as well as the *wbm_hit_7D* sequence, as a query, we performed a similarity search against de novo genome assembly of 10 wheat cultivars. We found one wheat cultivar, Mace, which shows absolute similarity with *wbm* and its promoter sequence, whereas no absolute similarity with *wbm-like* was observed, pointing to the absence of *wbm-like* in the genome of this cultivar and presence of *wbm*. The similarity search for *wbm_hit_7D* resulted in a hit with 100% similarity in the Mace cultivar located on distinct chromosome from *wbm* gene hit, further supporting that *wbm_hit_7D* is less related to *wbm* gene than *wbm-like*. Thus, the obtained results show that *wbm* and *wbm-like* are orthologous and located in the long arm of chromosome 7A.

3.2. *Comparative Analysis of wbm and wbm*-Like

To trace the origin of *wbm* and *wbm-like* (cultivar CS), we conducted multiple alignment of their nucleotide and protein sequences with sequences of *Triticum urartu* (A genome donor), *Aegilops speltoides* (B genome donor), *A. tauschii* (D genome donor), *T. monococcum*, and *Secale cereal*, followed by phylogenetic tree construction. *wbm_hit_7D*, located on chromosome 7D, and *wbm* sequences from two additional cultivars (Mace, which has *wbm*, and Julius, which has *wbm-like*) were also involved in the multiple sequence alignment. Comparison of the nucleotide sequences resulted in distinct cluster generated by the reference and Mace *wbm* genes (Cluster 1) and three clusters formed by *S. cereale* (Cluster 2), *A. speltoides* (Cluster 3), and *wbm-like* sequences from other species (Cluster 4) (Figure 1A,C). We next assess whether the comparison of translated protein sequences will show similar phylogenetic picture. Prediction of protein sequences from *wbm_hit_7D* and *A. tauschii* revealed a premature termination codon at the same position, indicating that sequences may not encode proteins.

Multiple alignment of all other predicted proteins (Figure 1B) and phylogenetic tree construction resulted in three clusters, generated by *wbm* (Cluster 1), *wbm-like* (Cluster 2), and *wbm* similar sequences from *S. cereale*, *T. urartu*, *A. speltoides*, and *T. monococcum* (Cluster 3). Thus, both phylogenetic trees show the distinct phylogenetic position of *wbm* gene. Interestingly, with a high divergence rate between the sequences from different clusters, N-terminus, corresponding to the signal peptide of the *wbm* protein [11], is well conserved.

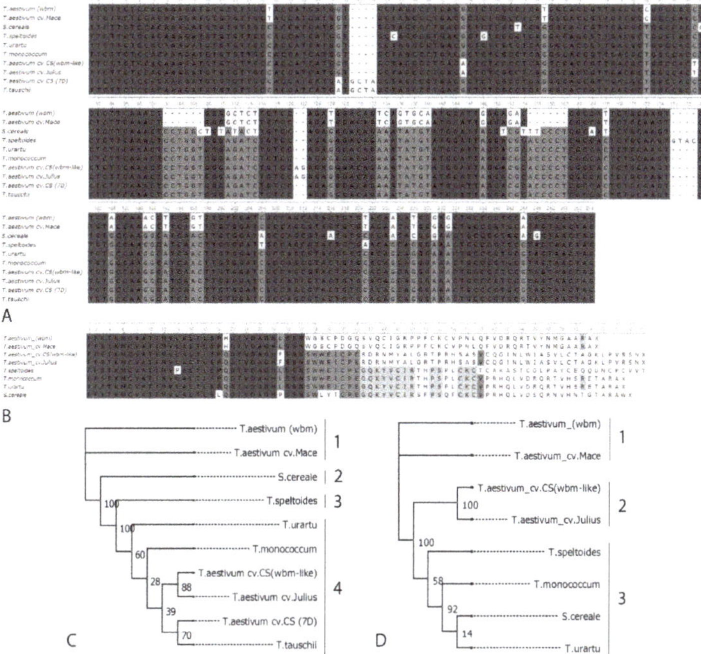

Figure 1. Conservation analysis of *wbm* and *wbm-like*. Multiple alignment of *wbm* and *wbm-like* (**A**) nucleotide and (**B**) amino acid sequences with similar sequences from *Triticum urartu*, *Aegilops speltoides*, *Triticum monococcum*, *Secale cereale*, and *Aegilops tauschii*. (**C**) Phylogenetic tree based on multiple alignment of *wbm(-like)* (**C**) nucleotide and (**D**) amino acid sequences. Bootstrap values are indicated.

Taken together, the comparative analysis indicated that *wbm* has a distinct origin from wheat genome donor species, supporting its introgressive origin in wheat. In addition, *wbm*, *wbm-like*, and *wbm* similar sequences from analyzed species demonstrate signatures of diversifying selection with high conservation in the N-terminus of protein.

3.3. *Screening of Triticale Collection in the Presence of wbm*

The established location of *wbm* and *wbm-like* on the chromosome 7A (Figure 2A) provided the possibility of checking for the presence of the genes in triticale possessing A, B, and R genomes but not D genome. The previously designed PCR marker (NWP, [11]) on *wbm* was dominant, resulting in amplification only in *wbm*-carrying genotypes. Therefore, we designed new *wbm-like*-specific primers on the upstream region (pro primers, Figure 2B) and CDS (*wbm-like* primers, Figure 2B) of the gene (Figure 2). In addition, *wbm*-specific primers on the CDS region of the gene were also designed (*wbm* primers, Figure 2B). In this study, we used the triticale germplasm collection consisting of 107 lines, most of which were obtained in Russia (Supplementary Table S1). Using the primer sets, we conducted PCR screening of this collection. Screening of this collection with *wbm* specific primers (NWP and *wbm*) revealed amplification of the PCR product of expected size (961 bp for NWP and 988 for pro, Figure 2C) only in three triticale lines, L8665, P13-5-13, and P13-5-2. The results of PCR screening with *wbm-like* specific primers showed that the expected PCR products were obtained for all lines except three (L8665, P13-5-13, and P13-5-2; Figure 2D), which do not possess *wbm-like*. Thus, we identified three triticale lines carrying *wbm* that can be further used in plant breeding programs to improve the bread-making quality of triticale.

3.4. Genomic Region of *wbm*-Like Lacks Conservation with *wbm* and Surrounded by Transposon Insertions

To understand the conservation rate of the genomic regions near *wbm* and *wbm-like*, promoter sequences (~−1 Kb region) of the genes were sequenced. Comparison of the obtained sequences demonstrated only partial similarity of the 67 bp region upstream of *wbm* and *wbm-like*. Analysis of the *wbm* genome locus in the CS genome assembly showed that the gene is surrounded by two transposable element (TE) insertions, and none of them were observed in the promoter of *wbm*. Of them, the TE located downstream (ca. 350 bp) belong to mutator DNA transposons family, whereas the TE located ca. 270 bp upstream of *wbm* (promoter) is L1-like retrotransposon (L1). To verify that the presence of L1 elements is unique to lines carrying *wbm-like*, we designed a primer pair (*wbm*_L1) flanking the L1 insertion. The screening of the triticale collection with this primer pair showed that the PCR product of expected size was obtained only in lines possessing *wbm-like*; no amplification was observed with DNA of lines possessing *wbm* (Figure 2C). Comparison of the *wbm-like* promoter with the *T. urartu* genome resulted in the finding of a highly similar sequence that also contains LINE1 insertion.

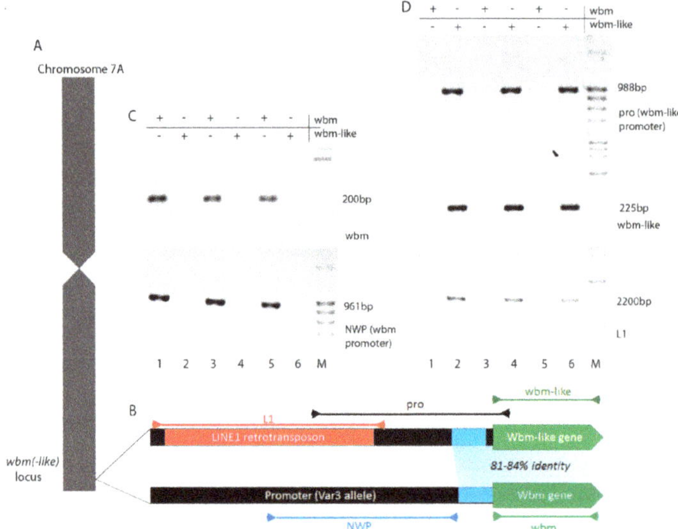

Figure 2. Genomic organization of *wbm-like* and *wbm* in genome of Chinese Spring. (**A**) Schematic representation of Chromosome 7A with marked position of *wbm(-like)* gene position. (**B**) Schematic organization of *wbm* and *wbm-like* and their promoter regions. The blue box depicts regions with high similarity (81% to 84%) and red box depicts Long interspersed nuclear element (LINE1) retroelement insertion. Horizontal lines show positions of primers. Gel electrophoresis of PCR products obtained with DNA of triticale lines (6 lines are represented here), and the primers specifically designed for (**C**) *wbm* gene (NWP (Furtado et al. [11]), *wbm* (specific for *wbm* open reading frame (ORF))) and (**D**) *wbm-like* gene (pro (promoter region of *wbm-like*), L1 (flanking LINE1 insertion in promoter region of *wbm-like*), and *wbm-like* (specific for *wbm-like* ORF)). 1–6 lanes of the gels corresponding to the 3 *wbm* positive lines (lane 1: L8665, lane 3: P13-5-13, and lane 5: P13-5-2) and 3 randomly selected *wbm-like* positive lines (lane 2: 131/7, lane 4: C 235, and lane 6: Yarilo).

The insertion of transposable element nearby genes often results in silencing of the gene expression. The expression of *wbm-like* in the CS cultivar was almost undetectable (wheat-expression.com).

Thus, the results show that upstream (promotor) regions of *wbm* and *wbm*-like genes have substantial differences supporting the introgressive origin of *wbm* gene. Further studies are needed to establish the borders and the size of the introgression.

4. Discussion and Conclusions

Genomes of modern wheat cultivars and landraces are shaped by inter- and intra-species introgression events [31]. The functional consequences of the introgressions and their effect on phenotypic variation and traits beneficial to humans, such as bread-making quality, have been poorly explored. Our study showed that *wbm* is a part of genomic introgression of unknown scale into the long arm of chromosome 7A. The protein and nucleotide sequences of *wbm* are phylogenetically distant from the homologous sequence of the A genome donor species, pointing to its distinct origin from unknown species. Concluding whether the introgression was a result of interspecies hybridization during the wheat breeding process or if it existed before the origin of wheat is difficult. Furtado et al. [11] reported the presence of *wbm* only in some genotypes of *T. monococcum* and *T. urartu*, suggesting that both scenarios are possible. With such a high divergence rate of *wbm* locus, the resequencing and read mapping to the reference genome (CS cultivar) will not allow identifying the origin of *wbm*, as it requires substantial similarity between reference and target genomes. Therefore, ongoing de novo sequencing and genome assembly of wheat cultivars (10+ Genome Project, http://www.10wheatgenomes.com/) and closely-related species are essential to identify the introgression similar to *wbm*.

The function of *wbm* protein remains elusive. Based on the specific spatiotemporal expression pattern and cysteine content of *wbm* protein, Furtado et al. [11] hypothesized that *wbm* protein can influence breadmaking quality through the generation of inter and intramolecular disulfide bonds. Notably, *wbm* protein is encoded by a small open reading frame (sORF). sORFs underwent rapid genomic evolution and may occur de novo in the genomes [32]. Comparison of the *wbm* protein sequence demonstrated high divergence between closely related species, consistent with general patterns of sORF evolution. The length of putative *wbm*-like proteins from the CS cultivar and closely related species also differ from the *wbm* protein. Although the molecular mechanism of *wbm* function is unknown, whether the sequence divergence is coupled with changes in *wbm* function is a topic that requires future research.

The established location of *wbm* on the A genome enabled the screening of a collection of triticale genotypes possessing A, B, and R genomes, and we found only three lines possessing *wbm*. Although a number of desirable traits in triticale have been inherited from both parents [13,14], the genes controlling desirable dough properties are located on the D genome and therefore are not present in hexaploid triticale [14]. Hence, new genetic resources are required to produce triticale producing strong gluten with high extensibility. Storage proteins of triticale, secaloglutenins, include a combination of HMW and LMW glutenins from the A and B genomes, secalins from the R genome, and α-, β-, and γ-gliadins [33,34]. Triticale lines possessing *wbm* may have unique gluten properties if *wbm* is capable of generating links between different storage proteins. However, further testing of this hypothesis requires a substantial amount of seeds to perform the baking test for these lines, which is the goal for our future research.

Here, we showed that GWseqVar3, previously described as an allelic variant of *wbm* [11], is orthologous to *wbm-like* located on chromosome 7A of wheat. In turn, *wbm-like* the CS cultivar is most probably a pseudogene surrounded by insertion of mobile elements, which has a negligible expression in grain. We found that *wbm* previously introgressed into the genomes of some wheat and triticale cultivars from unknown species. Elucidating the evolutionary relationships between *wbm* and its homologous sequences in wheat and closely related species allowed us to demonstrate that *wbm* protein is highly diverged, except its N-terminal. Finally, the location of *wbm* in the A genome offered the opportunity to select three triticale lines carrying this gene that can be further exploited in the future for improving the bread-making quality of this crop.

Supplementary Materials: The following are available online at http://www.mdpi.com/2073-4395/9/12/854/s1, Table S1: the list of triticale genotypes used in this study.

Author Contributions: Conceptualization, I.K. and A.S.; methodology, I.K.; formal analysis, A.P., N.M., M.D., G.K., M.K., I.G., and S.S.; writing—original draft preparation, I.K.; writing—review and editing, I.K., L.K., and A.S.; funding acquisition, A.S., G.K.

Funding: This research was funded by the Ministry of Education and Science of Russian Federation (goszadanie № 0431-2019-0005).

Conflicts of Interest: The authors declare no conflicts of interest.

References

1. Guzmán, C.; Xiao, Y.; Crossa, J.; González-Santoyo, H.; Huerta, J.; Singh, R.; Dreisigacker, S. Sources of the highly expressed wheat bread making (*wbm*) gene in CIMMYT spring wheat germplasm and its effect on processing and bread-making quality. *Euphytica* **2016**, *209*, 689–692. [CrossRef]
2. Payne, P.I.; Nightingale, M.A.; Krattiger, A.F.; Holt, L.M. The relationship between HMW glutenin subunit composition and the bread-making quality of British-grown wheat varieties. *J. Sci. Food Agric.* **1987**, *40*, 51–65. [CrossRef]
3. Morris, C.F. Puroindolines: The molecular genetic basis of wheat grain hardness. *Plant Mol. Biol.* **2002**, *48*, 633–647. [CrossRef] [PubMed]
4. Guzmán, C.; Alvarez, J.B. Wheat waxy proteins: Polymorphism, molecular characterization and effects on starch properties. *Theor. Appl. Genet.* **2016**, *129*, 1–16. [CrossRef] [PubMed]
5. D'Ovidio, R.; Masci, S.; Porceddu, E. Development of a set of oligonucleotide primers specific for genes at the Glu-1 complex loci of wheat. *Theor. Appl. Genet.* **1995**, *91*, 189–194. [CrossRef] [PubMed]
6. Lafiandra, D.; Tucci, G.; Pavoni, A.; Turchetta, T.; Margiotta, B. PCR analysis of x-and y-type genes present at the complex Glu-A1 locus in durum and bread wheat. *Theor. Appl. Genet.* **1997**, *94*, 235–240. [CrossRef]
7. Ma, W.; Zhang, W.; Gale, K. Multiplex-PCR typing of high molecular weight glutenin alleles in wheat. *Euphytica* **2003**, *134*, 51–60. [CrossRef]
8. Liang, X.; Zhen, S.; Han, C.; Wang, C.; Li, X.; Ma, W.; Yan, Y. Molecular characterization and marker development for hexaploid wheat-specific HMW glutenin subunit 1By18 gene. *Mol. Breed.* **2015**, *35*, 221. [CrossRef]
9. Kiszonas, A.M.; Morris, C.F. Wheat breeding for quality: A historical review. *Cereal Chem.* **2018**, *95*, 17–34. [CrossRef]
10. Rasheed, A.; Jin, H.; Xiao, Y.; Zhang, Y.; Hao, Y.; Zhang, Y.; Hickey, L.T.; Morgounov, A.I.; Xia, X.; He, Z. Allelic effects and variations for key bread-making quality genes in bread wheat using high-throughput molecular markers. *J. Cereal Sci.* **2019**, *85*, 305–309. [CrossRef]
11. Furtado, A.; Bundock, P.C.; Banks, P.; Fox, G.; Yin, X.; Henry, R. A novel highly differentially expressed gene in wheat endosperm associated with bread quality. *Sci. Rep.* **2015**, *5*, 10446. [CrossRef] [PubMed]
12. Jessop, R.S. Stress tolerance in newer triticales compared to other cereals. In *Triticale: Today and Tomorrow*; Springer: Berlin/Heidelberg, Germany, 1996; pp. 419–427.
13. Furman, B.J.; Qualset, C.O.; Skovmand, B.; Heaton, J.H.; Corke, H.; Wesenberg, D.M. Characterization and analysis of North American triticale genetic resources. *Crop Sci.* **1997**, *37*, 1951–1959. [CrossRef]
14. Dennett, A.L.; Cooper, K.V.; Trethowan, R.M. The genotypic and phenotypic interaction of wheat and rye storage proteins in primary triticale. *Euphytica* **2013**, *194*, 235–242. [CrossRef]
15. Blum, A. The abiotic stress response and adaptation of triticale—A review. *Cereal Res. Commun.* **2014**, *42*, 359–375. [CrossRef]
16. Randhawa, H.; Bona, L.; Graf, R. Triticale breeding—Progress and prospect. In *Triticale*; Springer: Berlin/Heidelberg, Germany, 2015; pp. 15–32.
17. Liu, W.; Maurer, H.P.; Leiser, W.L.; Tucker, M.R.; Weissmann, S.; Hahn, V.; Würschum, T. Potential for marker-assisted simultaneous improvement of grain and biomass yield in triticale. *Bioenergy Res.* **2017**, *10*, 449–455. [CrossRef]

18. Ayalew, H.; Kumssa, T.T.; Butler, T.J.; Ma, X.-F. Triticale Improvement for Forage and Cover Crop Uses in the Southern Great Plains of the United States. *Front. Plant Sci.* **2018**, *9*, 1130. [CrossRef]
19. Woś, H.; Brzeziński, W. Triticale for food—The quality driver. In *Triticale*; Springer: Berlin/Heidelberg, Germany, 2015; pp. 213–232.
20. McGoverin, C.M.; Snyders, F.; Muller, N.; Botes, W.; Fox, G.; Manley, M. A review of triticale uses and the effect of growth environment on grain quality. *J. Sci. Food Agric.* **2011**, *91*, 1155–1165. [CrossRef]
21. Consortium, I.W.G.S. Shifting the limits in wheat research and breeding using a fully annotated reference genome. *Science* **2018**, *361*, eaar7191.
22. Alaux, M.; Rogers, J.; Letellier, T.; Flores, R.; Alfama, F.; Pommier, C.; Mohellibi, N.; Durand, S.; Kimmel, E.; Michotey, C.; et al. Linking the International Wheat Genome Sequencing Consortium bread wheat reference genome sequence to wheat genetic and phenomic data. *Genome Biol.* **2018**, *19*, 111. [CrossRef]
23. Altschul, S.F.; Gish, W.; Miller, W.; Myers, E.W.; Lipman, D.J. Basic local alignment search tool. *J. Mol. Biol.* **1990**, *215*, 403–410. [CrossRef]
24. Bian, X.; Tyrrell, S.; Davey, R.P. The Grassroots Life Science Data Infrastructure (2017). Available online: https://grassroots.tools (accessed on 1 September 2019).
25. Howe, K.L.; Contreras-Moreira, B.; De Silva, N.; Maslen, G.; Akanni, W.; Allen, J.; Alvarez-Jarreta, J.; Barba, M.; Bolser, D.M.; Cambell, L. Ensembl Genomes 2020—Enabling non-vertebrate genomic research. *Nucleic Acids Res.* **2019**. [CrossRef]
26. Katoh, K.; Standley, D.M. MAFFT multiple sequence alignment software version 7: Improvements in performance and usability. *Mol. Biol. Evol.* **2013**, *30*, 772–780. [CrossRef] [PubMed]
27. Okonechnikov, K.; Golosova, O.; Fursov, M.; Team, U. Unipro UGENE: A unified bioinformatics toolkit. *Bioinformatics* **2012**, *28*, 1166–1167. [CrossRef] [PubMed]
28. Ramírez-González, R.; Borrill, P.; Lang, D.; Harrington, S.; Brinton, J.; Venturini, L.; Davey, M.; Jacobs, J.; Van Ex, F.; Pasha, A. The transcriptional landscape of polyploid wheat. *Science* **2018**, *361*, eaar6089. [CrossRef] [PubMed]
29. Wickham, H. *ggplot2: Elegant Graphics for Data Analysis*; Springer: Berlin/Heidelberg, Germany, 2016.
30. Untergasser, A.; Cutcutache, I.; Koressaar, T.; Ye, J.; Faircloth, B.C.; Remm, M.; Rozen, S.G. Primer3—New capabilities and interfaces. *Nucleic Acids Res.* **2012**, *40*, e115. [CrossRef]
31. He, F.; Pasam, R.; Shi, F.; Kant, S.; Keeble-Gagnere, G.; Kay, P.; Forrest, K.; Fritz, A.; Hucl, P.; Wiebe, K. Exome sequencing highlights the role of wild-relative introgression in shaping the adaptive landscape of the wheat genome. *Nat. Genet.* **2019**, *51*, 896. [CrossRef]
32. Fesenko, I.; Kirov, I.; Kniazev, A.; Khazigaleeva, R.; Lazarev, V.; Kharlampieva, D.; Grafskaia, E.; Zgoda, V.; Butenko, I.; Arapidi, G. Distinct types of short open reading frames are translated in plant cells. *Genome Res.* **2019**, *29*, 1464–1477. [CrossRef]
33. Salmanowicz, B.P.; Dylewicz, M. Identification and characterization of high-molecular-weight glutenin genes in Polish triticale cultivars by PCR-based DNA markers. *J. Appl. Genet.* **2007**, *48*, 347–357. [CrossRef]
34. Amiour, N.; Bouguennec, A.; Marcoz, C.; Sourdille, P.; Bourgoin, M.; Khelifi, D.; Branlard, G. Diversity of seven glutenin and secalin loci within triticale cultivars grown in Europe. *Euphytica* **2002**, *123*, 295–305. [CrossRef]

Communication

Recovery of 2R.2S^{k} Triticale-*Aegilops kotschyi* Robertsonian Chromosome Translocations

Waldemar Ulaszewski [1], Jolanta Belter [1], Halina Wiśniewska [1], Joanna Szymczak [2], Roksana Skowrońska [2], Dylan Phillips [3] and Michał T. Kwiatek [1,*]

1 Department of Genomics, Institute of Plant Genetics of the Polish Academy of Sciences, Strzeszyńska 34, 60-479 Poznań, Poland; wula@igr.poznan.pl (W.U.); jbel@igr.poznan.pl (J.B.); hwis@igr.poznan.pl (H.W.)

2 Department of Genetics and Plant Breeding, Poznań University of Life Sciences, Dojazd 11, 60-632 Poznań, Poland; asia.szymczak28@wp.pl (J.S.); roksana.skowronska@up.poznan.pl (R.S.)

3 Institute of Biological, Environmental and Rural Sciences, Aberystwyth University, Aberystwyth, Ceredigion, Wales SY23 3DA, UK; dwp@aber.ac.uk

* Correspondence: michal.kwiatek@up.poznan.pl; Tel.: +48-61-848-7760

Received: 9 September 2019; Accepted: 15 October 2019; Published: 17 October 2019

Abstract: Robertsonian translocations (RobTs) in the progeny of triticale (×*Triticosecale* Wittmack) plants with monosomic substitution of *Aegilops kotschyi* chromosome 2S^{k} (2R) were investigated by fluorescence in-situ hybridization. Chromosome 2S^{k} of *Ae. kotschyi* is reported to possess many valuable loci, such as *Lr54* + *Yr37* leaf and stripe (yellow) rust resistance genes. We used a standard procedure to produce RobTs, which consisted of self-pollination of monosomic triticale plants, carrying 2R and 2S^{k} chromosomes in monosomic condition. This approach did not result in RobTs. Simultaneously, we succeeded in producing 11 plants carrying 2R.2S^{k} compensatory RobTs using an alternative approach that utilized ditelosomic lines of triticale carrying 2RS (short arm) and 2RL (long arm) telosomic chromosomes. Identification of molecular markers linked to *Lr54* + *Yr37* genes in the translocation plants confirmed that these resources can be exploited in current triticale breeding programmes.

Keywords: *Aegilops*; centric breaks; chromosome fusion; Robertsonian translocations; telosomic chromosomes; triticale

---

1. Introduction

Hexaploid triticale (×*Triticosecale* Wittmack, $2n = 6x = 42$, AABBRR) is one of the few artificial crops cultivated at a large scale. Triticale combines the quality traits of wheat (*Triticum aestivum* L.) and the adaptation abilities of rye (*Secale cereale* L). Over the last three decades the global harvested area of triticale has constantly increased (2,101,405 ha in 1996; 3,662,363 ha in 2006, and 4,157,018 ha in 2016) [1], and the range of uses has also grown. Forage production is the principal end use for this crop, but there are new niches proposed, such as: biofuel production [2,3], baking [4], brewing [5] and food production [6]. As a recent man-made crop, triticale suffers narrow genetic variability for breeders to select upon.

In triticale there has been a breakdown of resistance against leaf rust (caused by *Puccinia triticiana* Eriks.) and stripe (yellow) rust (caused by *P. striiformis* f. sp. *tritici* Westend.). As the harvested area of triticale has increased, new pathotypes of *Puccinia* have evolved, moving from wheat and rye into triticale [7]. Both pathogens can reduce the grain yield by 40% [8,9]. There is an urgent need to improve the genepool of triticale, and introduce genetic resistance to *Puccinia* infections. There are approximately eighty leaf rust resistance genes identified in Triticeae, and nearly the same amount of stripe rust resistance genes have also been identified. Some of these genes have already been transferred from wild relatives into the wheat genetic background [10]. Recently, several attempts were made to transfer rust resistance genes from *Aegilops*, *Agropyron* and *Triticum* species into triticale [11–15].

Aegilops species are closely related to wheat (and triticale, *per se*) and carry a number of valuable traits, which have been effectively incorporated into wheat by developing wheat–*Aegilops* hybrids and deriving addition, substitution and translocation lines [16]. *Aegilops kotschyi* Boiss. ($2n = 4x = 28$ chromosomes, U- and S-genomes) is a wild tetraploid goatgrass native to Northern Africa, the Mid-East, and Western Asia. *Ae. kotschyi* germplasm is exploited in wheat breeding [17] as a source of high grain protein, iron and zinc [18]. Moreover, Antonov and Marais [19] observed leaf rust resistance that was effective against the infection of *Puccinia triticina* in *Ae. kotschyi*. Marais et al. [20] identified the *Lr54* and *Yr37* leaf rust and stripe rust resistance genes, and developed aT2DS.2S^{k}L wheat-*Ae. kotschyi* translocation line. The first *Lr54* + *Yr37* marker was developed by Heyns et al. [21]. Moreover, translocation gene sequences were cloned and specific SSR markers were developed [22].

Homoeologous recombination based engineering is the most common way for efficiently utilizing the wild relative gene pool for crop improvement [23]. The generation of translocation lines is the most promising pathway for the exploitation of alien germplasm in crop breeding [23]. In distant hybrids, unpaired chromosomes are present as univalents during meiosis. Monosomic chromosomes are prone to centric breaks at anaphase I of meiosis, which misdivide and the broken ends fuse during the interkinesis of meiosis II [24–26]. Fusion of the misdivided products may result in the formation of a Robertsonian translocation (RobT) [27].

Several steps are required to generate RobTs (Figure 1a), with self-pollination of double-monosomic plants being the most common method used in the induction of RobTs [26]. Wheat breeders can use a large collection of aneuploid stocks for the induction of wheat-wheat and wheat-alien RobTs. This can be performed in a directed manner by producing the appropriate wheat or alien monosomic lines [28–31]. In wheat breeding the most common RobTs are the T1BL.1RS and T1AL.1RS translocations [25]. Lukaszewski and Curtis developed 1RS.1DL chromosome translocation in triticale "Rhino" and transferred this to cv. "Presto" by backcrossing for several generations [32]. The 1RS.1DL translocation was later used for the induction of multi-breakpoint translocation lines [29,33,34]. A key issue for the centric breakage-fusion mechanism is the frequency at which the centric breakage events involving the univalent occurs. Friebe et al. [26] reported that this can range from 1% to 11%. Another issue is the resultant telocentric chromosomes have a tendency of rejoin as RobTs. It was reported that the frequency of wheat–alien RobTs recovered ranges from 4% to 20%, depending on the genetic background, the chromosomes involved, and environmental conditions used [26,30,31].

To increase the rate of RobT formation, it is therefore necessary to reduce the random factors connected with appearance of centric breaks. In this study, we postulated that the use of telosomic plants for cross-hybridization with plants carrying an alien-substitution will overcome the random process of centric break formation in the univalent (Figure 1b). This approach was used to transfer the *Lr54* + *Yr37* resistance genes into triticale cv. "Sekundo" through a T2RS.2S^{k}L RobT. It was hypothesized that the presence of the 2R donor chromosomes in a telosomic condition would increase the frequency of RobTs recovery (Figure 1b). To test this we used a monosomic substitution triticale line carrying a single 2S^k chromosome, instead of a 2R triticale chromosome (40 + M2R + M2S^k), and crossed this with a triticale line carrying a single copy of 2RS (short arm of 2R chromosome) and 2RL (long arm of 2R chromosome) telosomic chromosomes (40T + D2RS + D2RL). As a control experiment, the classical approach, based on the self-pollination of double monosomic triticale-*Ae. kotschyi* plants (40T + M2R + M2S^k), was also undertaken (Figure 1a). The goal was to compare the frequencies of chromosomal breaks and recovery of RobTs that arose throughout the two independent approaches tested (Figure 1a,b).

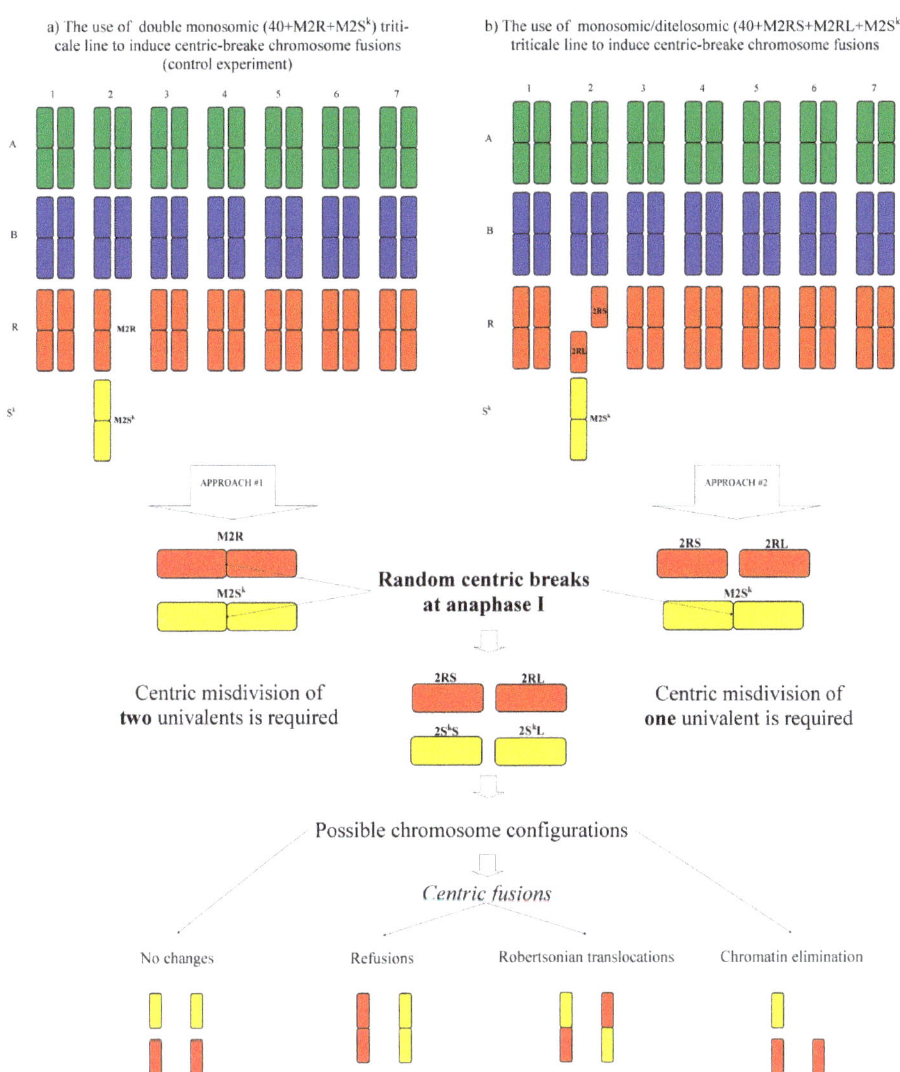

Figure 1. Ideogram of the two contrasting approaches for the recovery of Robertsonian translocations.

2. Materials and Methods

2.1. Plant Material

Winter triticale cv. "Sekundo" was obtained from the Danko Hodowla Roślin Sp. z o.o. (Choryń, Poland) plant breeding company. Accession Ak3 of *Aegilops kotschyi* was received by Prof. B. Wojciechowska (Institute of Plant Genetics of the Polish Academy of Sciences, Poznań, Poland; IPG PAS) from a collection of Professor M. Feldman (Weizmann Institute of Science, Rehovot, Israel) and assembled in the genebank of the IPG PAS. All plants were grown and crossed in the greenhouse of the IPG PAS. In the initial study, monosomic addition triticale line (MA2S^k; $2n$ = 43) that carried chromosome 2S^k derived from triticale-*Aegilops kotschyi* hybrids was used. These plants were crossed with triticale nullisomic plants for chromosomes 2R (N2R, $2n$ = 40); obtained by Kwiatek et al. [35]. Double monosomic (M2R and M2S^k) F_1 plants ($2n = 42$) were identified using GISH, according to Kwiatek et al. [35]. Ditelosomic (Dt2RS and Dt2RL) plants of triticale were generated by Kwiatek et al. [35] by harnessing chromosome fragmentation induced by gametocidal action, in combination with the production of double haploids. Spikes emasculated between April and June of 2018 were used for cross-pollination. The anthers of donor plants were collected, and pollen was transferred onto stigmas by chafing.

2.2. Chromosome Preparation

Accumulation of cells at mitotic metaphase and fixation was carried out according to Kwiatek et al. [35]. The root meristemes were digested at 37 °C for 2 h and 40 min in an enzymes solution containing 0.2% (*v*/*v*) Cellulase Onozuka R-10 and Calbiochem cytohelicase (1:1 ratio) and 20% pectinase (Sigma), in 10 mM citrate buffer (pH 4.6). Chromosome preparations were made according Heckmann et al. [36]. Digested root tips were placed on slides with a drop of ice-cold 60% acetic acid. The mixture was spread on a slide using metal needle for 2 min on a heating table (Medax) set at 48 °C. The slides were washed with 200 μL of ice-cold ethanol-acetic acid (3:1, *v*/*v*) and placed in 60% acetic acid for 10 min, washed in 96% ethanol, and air dried.

Meiotic chromosome spreads from pollen mother cells were prepared from flower buds fixed with ethanol-acetic acid (3:1, *v*/*v*) according to Zwierzykowski et al. [37]. Anthers were transferred to a slide with a drop of ice-cold 60% acetic acid and dispersed with a metal needle. The slide was placed on a hot plate (45 °C) for 2 min, during which the drop was spread using a needle. The slide was removed from the hot plate, a drop of 60% acetic acid added and covered with a cover slip. The slide was frozen in liquid nitrogen, and cover slip was removed with a razor blade.

2.3. Probe Preparation and Fluorescence in Situ Hybridization

Total genomic DNA was isolated using GeneMATRIX Plant & Fungi DNA Purification Kit (EURx, Gdansk, Poland). DNA of *Aegilops sharonensis* Eig. (a progenitor of the S-genome of *Ae. kotschyi*; PI 551020, U.S. National Plant Germplasm System, Aberdeen, ID, United States of America) was labeled by nick translation with Atto-488 dye (Atto-488NT kit; Jena Bioscience, Jena, Germany) to investigate *Aegilops* chromosomes behavior during meiosis in the hybrid. Blocking DNA from triticale (Sekundo, Lamberto and Bogo) was sheared by boiling for 30-45 min and used at a ratio of 1:50 (probe:block). Genomic in-situ hybridization (GISH) or multicolor GISH was carried out according to previously published protocols [35]. Mitotic chromosomes were identified using fluorescence in situ hybridization (FISH) with the repetitive sequences from pTa-86, pTa-535, pTa-465, pTa-k-566 and pTa713 clones characterized by Komuro et al. [38]. They were amplified from genomic DNA of wheat (Chinese Spring) according to Kwiatek et al. [39], and labelled with Atto-488, Atto-550, and Atto-647 dyes using a nick translation kit (Jena Bioscience). FISH was performed according to Kwiatek et al. [35]. Slides were examined with the Olympus BX 61 automatic epifluorescence microscope equipped with Olympus XM10 CCD camera. Image processing was carried out using Olympus Cell-F (version 3.1; Olympus Soft Imaging Solutions GmbH: Münster, Düsseldorf, Germany) imaging software and PaintShop Pro X5 software (version 15.0.0.183; Corel Corporation, Ottawa, ON, Canada). Chromosomes of *Aegilops* and triticale were identified by comparing the signal patterns of the probes [39].

2.4. Lr54 + Lr37 SSR Marker Screening

Genomic DNA of parental forms and offspring plants were isolated using Plant DNA Purification Kit (EurX Ltd., Gdańsk, Poland). All primers (Table 1) were manufactured by Sigma-Aldrich (Merck).

PCR reactions were performed in a LabCycler thermal cycler (SensoQuest Biomedizinische Elektronik, Goettingen, Germany). The 20 μL PCR reaction consisted of 150 nM each primer, 0.2 mM of each nucleotide, 1.5 mM $MgCl_2$, 0.2 units of Taq-DNA hot-start polymerase (TaqNovaHS, Blirt, Poland), and 50 ng of genomic DNA as a template. A typical PCR procedure was as follows: 5 min at 95 °C, then 35 cycles of 30 s at 94 °C, 30 s at 50–60 °C (depending on the primer, Table 1), 1 min at 72 °C, and 5 min at 72 °C. 0.5µL Midori Green Direct (Nippon Genetics Europe) was added to each amplification product, ran on 2% agarose gel (Sigma), and then visualized and photographed (EZ GelDoc System, BioRad).

Table 1. Primer sequences and PCR conditions used for *Lr54* + *Lr37* marker identification on $2S^k$ chromosome.

Marker	Primer Sequence (5′ to 3′)	Amplification Temperature (°C)	Amplicon Size (bp)	Reference
S14-297	CATGCAGAAAACGACACACC GGTAAGTGGTCAGGCGTTGT	60	297	[21]
S14-410	ACCAATTCAACTTGCCAAGAG GAGTAACATGCAGAAAACGACA	61	410	[22]

3. Results

3.1. Chromosome Segregation in 40 + M2R + M2S^k Plants of Triticale

In the control experiment, pollen mother cells (PMCs) of ten triticale plants carrying 40 + M2R + M2S^k chromosomes were examined to identify metaphase I (MI) and anaphase I (AI) of meiosis (Figure 2a). We observed a 2S^k univalent in all of 100 analyzed PMCs (Figure 2b). The number of misdivided 2S^k chromosomes during AI was 9 (Figure 2c). No other misdivided chromosome of triticale was observed. Next, 48 spikes of the monosomic substitution plants (40 + M2R + M2S^k) were allowed to self-pollinate, and 2601 seeds were obtained. A subset of 289 plants were screened using GISH and none were found with the expected RobT. 185 (63.1%) of the derived plants carried a whole *Aegilops* chromosome. In the remaining offspring (104 plants; 36.9%), no *Ae. kotschyi* chromosome fragments were identified.

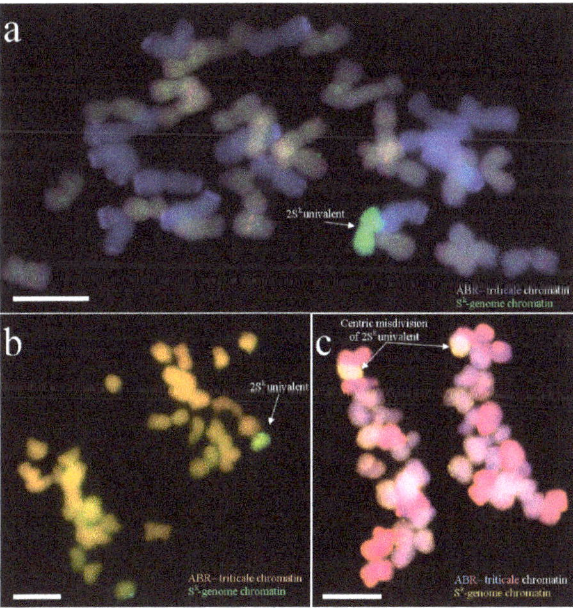

Figure 2. (**a**) Chromosome set of a 40 + M2R + M2S^k triticale plant (**b**) Segregation of chromosome 2S^k (green, Atto-488) at anaphase I in a 40 + M2R + M2S^k triticale plant. (**c**) Centric misdivision of chromosome 2S^k (yellow, Atto-647) at anaphase I in a 40 + M2R + M2S^k triticale plant. Scale bar, 10 µm.

3.2. *Chromosome Segregation in 40 + M2RS + M2RL + M2S^k Plants of Triticale*

The main goal of the experiment was to obtain triticale plants with introgression of a 2S^k chromosome segment on the 2R chromosome. Hence, we crossed monosomic substitution (40 + M2R + M2S^k) plants with double ditelosomic D2RSD2RL plants of triticale to obtain triple monosomic triticale plants (40 + M2RS + M2RL + M2S^k). These plants contained a single telosomic 2RS, a single telosomic 2RL and a single monosomic 2S^k chromosome (Figure 1b). 13,457 flowers of the ditelosomic line were pollinated with pollen collected from monosomic substitution M2S^k (M2R) plants, and 3464 seeds were obtained (25.74% crossing efficiency). The chromosomal constitution of 100 randomly chosen plants were examined by a combination of genomic and fluorescence in situ hybridization methods (GISH/FISH), using total genomic DNA of rye and *Aegilops* species, and pTa-103 centromere sequence, as probes. 38 plants were found to be triple monosomic (40 + M2RS + M2RL + M2S^k) (Figure 3), 35 lacked the 2R chromosomes (40 + N2R + M2S^k), 15 plants were ditelosomic (40 + M2RS + M2RL) and 12 plants were double telosomic (40 + D2RL).

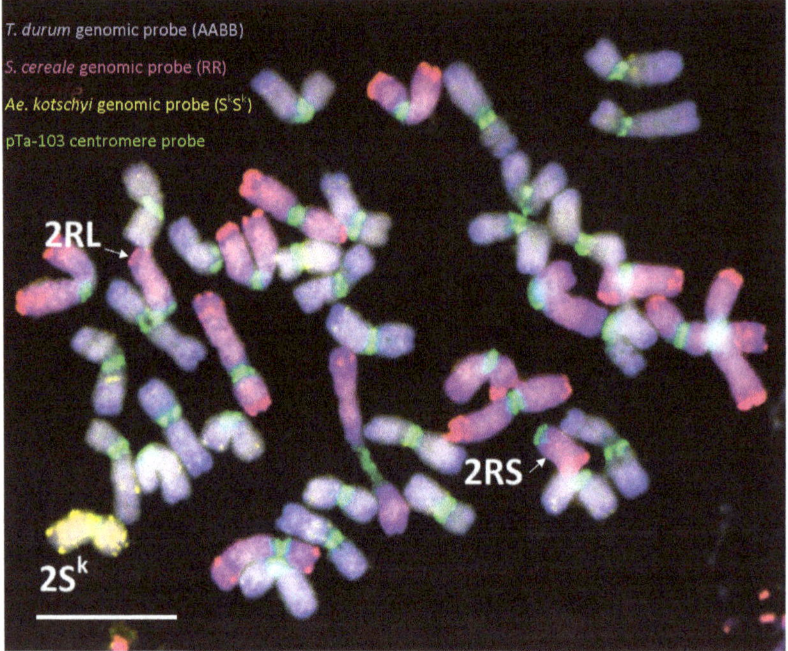

Figure 3. Karyotype of triticale plant carrying 40 + M2RS + M2RL + M2S^k chromosomes, examined by genomic/fluorescence in situ hybridization (GISH/FISH). Total genomic DNA of rye (red) and *Ae. tauschii* (yellow), and pTa103 centromere sequence (green) were used as probes. Scale bar, 10 µm.

The triple monosomics were selected from further analysis (38 plants; 40 + M2RS + M2RL + M2S^k), and MI and AI were observed in 10 PMCs of each plant. Chromosome 2S^k was present as a univalent in all PMCs examined. In most cases (35 plants; 92.11%), the *Aegilops* univalent underwent chromosome segregation and was moved randomly to one of the opposite poles of the cell (Figure 4a).

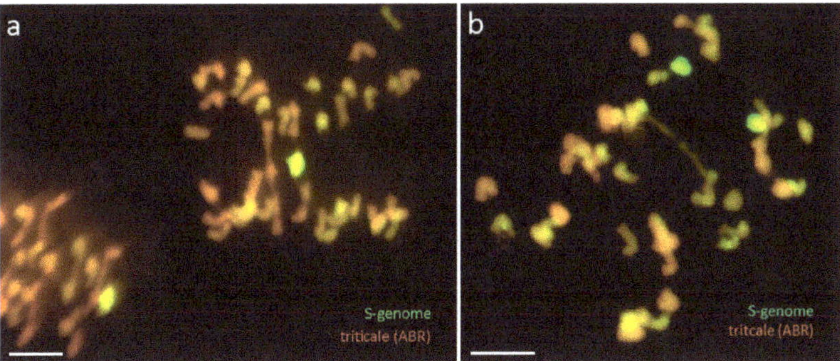

Figure 4. (**a**) Chromosome segregation of $2S^k$ (green, Atto-488) at anaphase I of triticale line 40 + M2RS + M2RL + $M2S^k$. (**b**) Centric misdivision of chromosome $2S^k$ (green, Atto-488) at anaphase I of triticale line 40 + M2RS + M2RL + $M2S^k$. Scale bar, 10 μm.

We also observed the misdivision of the $2S^k$ chromosome arms in three plants (7.89%) (Figure 4b). Next, a total number of 68 spikes of 38 offspring plants (40 + M2RS + M2RL + $M2S^k$) were self-pollinated, and 3489 seeds were obtained. The chromosome constitutions of 100 randomly chosen plants were examined by FISH/GISH, using repetitive sequences as probes (pTa-86; pTa-535 and total genomic DNA from rye). The results of chromosome $2S^k$ misdivision were observed in 13 plants, including: a double RobT-T2RS.$2S^k$L and T$2S^k$S.2RL (1 plant; Figure 5a, b and c); six plants carrying T2RS.$2S^k$L, four plants with T$2S^k$S.2RL, and two telosomic plants with the 2RS, 2RL and $2S^k$S chromosome arms (Table 2).

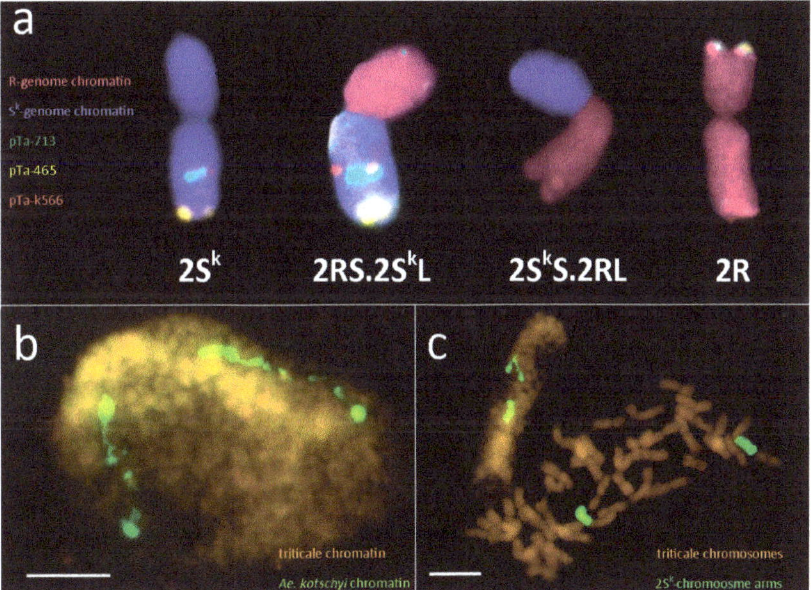

Figure 5. (**a**) T$2S^k$.2R Robertsonian translocations in triticale analysed by fluorescence in situ hybridization (R-genome chromatin—purple, S^k-genome chromatin—blue, pTa-713—green, pTa-465—yellow, pTa-k566—red). Genomic in-situ hybridization pattern of triticale plant carrying Robertsonian traslocations with segments of $2S^k$ chromosome (green) at prophase (**b**) and metaphase (**c**) of root meristem cell. Scale bar, 10 μm.

Table 2. Chromosome constitutions and Lr54 + Yr37 SSR marker analysis of triticale plants carrying divided $2S^k$ chromosome.

Plant No.	Chromosome Constitution	Chromosome Number	Amplicon Size for S14-410 Marker (bp)	Amplicon Size for S14-297 Marker (bp)
1	40 + M2RS + M2RL + M2S^{k}S	43	null	null
2	40 + T2RS.2S^{k}L + M2RL	42	410	297
3	40 + T2S^{k}S.2RL	41	null	null
4	40 + T2RS.2S^{k}L	41	410	297
5	40 + T2RS.2S^{k}L + M2RL	42	410	297
6	40 + T2RS.2S^{k}L + t2S^{k}S.2RL	42	410	297
7	40 + T2S^{k}S.2RL	41	null	null
8	40 + T2RS.2S^{k}L	41	410	297
9	40 + M2RS + M2RL + M2S^{k}S	43	null	null
10	40 + T2RS.2S^{k}L	41	410	297
11	40 + T2S^{k}S.2RL + M2RS	42	null	null
12	40 + T2RS.2S^{k}L + M2RL	42	410	297
13	40 + T2S^{k}S.2RL	41	null	null

3.3. Lr54 + Yr37 SSR Markers Analysis

Initial analysis of molecular markers S14-297 and S14-410, linked to Lr54 + Yr37 loci [21,22], was conducted on the parental lines; namely the monosomic substitution M2S^k (M2R) line and double ditelosomic D2RSD2RL triticale line. The same protocol was used to examine the subsequent offspring. An amplicon of 297bp (base pairs) was for obtained for the S14-297 marker, and 410bp for the S14-410 marker, as observed in control samples of *Ae. kotschyi* and double 40 + M2R + M2S^k monosomic plants (Figure 6). The same 410bp product was amplified form the 6 plants with T2RS.2S^{k}L RobTs and for 1 plant carrying double RobTs-T2RS.2S^{k}L and T2S^{k}S.2RL (Figure 6; Table 2).

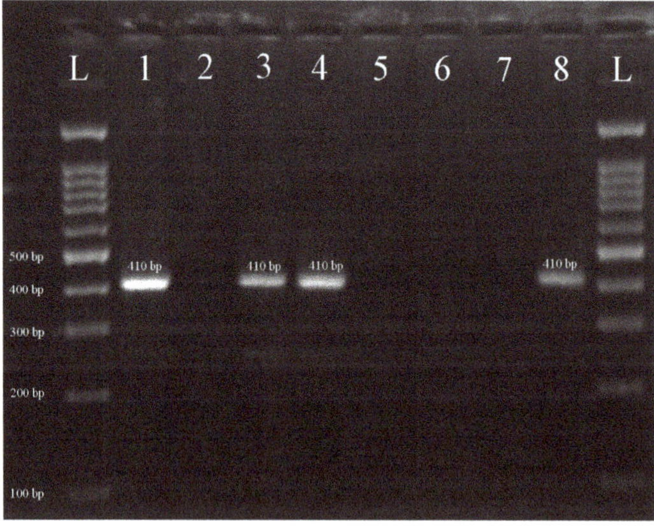

Figure 6. Amplification products of S14-410 marker linked to loci of *Lr54* + *Lr37* leaf and stem rust resistance genes. (**1**) *Aegilops kotschyi* (UUS^{k}S^k); (**2**) triticale 'Sekundo'(AABBRR); (**3**) double 40T + M2R + M2S^k monosomic; (**4**) T2RS.2S^{k}L and T2S^{k}S.2RL; (**5**) double ditelosomic D2RS + D2RL; (**6**) telosomic M2RS + M2RL + M2S^{k}S; (**7**) T2S^{k}S.2RL; (**8**) T2RS.2S^{k}L.

4. Discussion

Many reports have shown that RobTs in wheat can arise by centromeric misdivision of univalents at AI, followed by segregation of the derived telocentric chromosomes to the same nucleus, and fusion of the broken parts [26,30,31]. The focus of this study was to exploit the formation of RobTs to transfer of the *Lr54* + *Yr37* gene loci from *Ae. kotschyi* into triticale. We have focused on two essential aspects of this process: the frequency of univalent misdivision, and the frequency at which the broken chromosomes fuse to form a RobT. The control experiment used the standard approach of self-pollinating monosomic substitution (40 + M2R + M2S^k) of triticale. Genomic in-situ hybridization identified that chromosome 2S^k was present in all meiotic cells assayed, and the centromere was divided during AI in 9% of the PMCs surveyed. The same event was observed in PMCs isolated from the 40 + M2RS + M2RL + M2S^k plants. The frequency of PMCs where the centromere of the 2S^k univalent divided was 7.89%. Our results are comparable with other reports, showing that the frequency of univalent misdivision range between 3% up to 51% [26,30,31]. When both approaches were compared (40 + M2R + M2S^k vs 40 + M2RS + M2RL + M2S^k), no significant difference in the frequency at which the *Aegilops* univalent misdivided was observed. However, the 2R univalent did not misdivide in the control experiment, and the use of the double ditelosomic D2RS + D2RL in the alternative approach was justified.

It can be assumed that the probability of RobTs formation is dependent on the rate of centromere breakage, and subsequent formation of telocentric chromosomes, occurs during meiosis. In wheat, the frequency of recovery of wheat–alien RobTs is between 2% and 20%, depending on the chromosomes involved [23,26,30,31,40–44]. In our experiment, the progeny of monosomic substitution plants lacked any kind of triticale-*Aegilops* translocation. This could correlate with the lack of the 2R chromosome misdivision observed in these plants. Similar issues were reported by Ghazali et al. [45], who aimed to produce whole arm RobTs involving chromosomes 2B (wheat) and 2E^b (*Thinopyrum bessarabicum*). Surprisingly, the authors obtained only T2E^{b}S.2BL, while no 2BS.2E^{b}L RobTs were recovered [45]. It has been reported that rye chromosomes rearrange at a higher rate when compared to wheat chromosomes [46], with the long arm of chromosome 2R more readily fused with the chromosome arm of an alien species. For example, Fiebe et al. [47] derived 2BL.2RL RobTs carrying Hessian fly resistance gene *H21* in wheat cv. Hamlet. Rahmatov et al. [48] also obtained 2DS.2RL RobT, which transfered the stem rust resistance gene *Sr59* into wheat.

In our study, we have effectively used ditelosomic lines of triticale to increase the frequency of 2S^k.2R RobTs generated. Moreover, we have obtained both 2S^{k}S.2RL and 2RS.2S^{k}L RobTs. The ability to produce both types of RobTs was an important goal as the particular chromosome segments of *Aegilops* may harbor other desirable genes. To summarize, we obtained plants with chromosome translocations, which probably carry the *Lr54* + *Lr37* leaf and stem rust resistance genes. The next step of this project will be to phenotype the eleven RobTs lines for leaf rust or stripe rust resistance. If these tests succeed, such germplasm could be a valuable source for triticale resistance breeding.

Author Contributions: Conceptualization, M.T.K.; methodology, M.T.K.; validation, W.U.; formal analysis, W.U.; investigation, W.U., J.B., J.S. and R.S.; resources, M.T.K., H.W.; data curation, M.T.K., W.U.; writing—original draft preparation, W.U., M.T.K.; writing—review and editing, W.U., D.P. and M.T.K.; visualization, W.U., M.T.K.; supervision, M.T.K., D.P. and H.W.; project administration, M.T.K.; funding acquisition, M.T.K.

Funding: This research and the APC were funded by NATIONAL CENTRE FOR RESEARCH AND DEVELOPMENT, Poland (Narodowe Centrum Badań i Rozwoju, Polska), grant number LIDER/3/0004/L-8/16/NCBR/2017.

Acknowledgments: We would like to acknowledge the administrative support provided by the Institute of Plan Genetics of the Polish Academy of Sciences.

Conflicts of Interest: The authors declare no conflict of interest.

References

1. Food and Agriculture Organization of the United Nations. FAOSTAT Statistics Database 1998. Available online: http://www.fao.org/faostat (accessed on 28 June 2019).
2. McGoverin, C.M.; Snyders, F.; Muller, N.; Botes, W.; Fox, G.; Manley, M. A review of triticale uses and the effect of growth environment on grain quality. *J. Sci. Food Agric.* **2011**, *91*, 1155–1165. [CrossRef]
3. Meale, S.J.; McAllister, T.A. Grain for Feed and Energy. In *Triticale*; Eudes, F., Ed.; Springer International Publishing: Cham, Switzerland, 2015; pp. 167–187.
4. Pattison, A.L.; Trethowan, R.M. Characteristics of modern triticale quality: Commercially significant flour traits and cookie quality. *Crop Pasture Sci.* **2013**, *64*, 874–880. [CrossRef]
5. Rakha, A.; Åman, P.; Andersson, R. Dietary fiber in triticale grain: Variation in content, composition, and molecular weight distribution of extractable components. *J. Cereal Sci.* **2011**, *54*, 324–331. [CrossRef]
6. Woś, H.; Brzeziński, W. Triticale for Food—The Quality Driver. In *Triticale*; Eudes, F., Ed.; Springer International Publishing: Cham, Switzerland, 2015; pp. 213–232.
7. Arseniuk, E. Triticale Diseases—A Review. In *Triticale: Today and Tomorrow*; Guedes-Pinto, H., Darvey, N., Carnide, V.P., Eds.; Springer: Dordrecht, The Netherlands, 1996; pp. 499–525.
8. Arseniuk, E.; Góral, T. Triticale Biotic Stresses—Known and Novel Foes. In *Triticale*; Eudes, F., Ed.; Springer International Publishing: Cham, Switzerland, 2015; pp. 83–108.
9. Tian, S.; Weinert, J.; Wolf, G.A. Infection of triticale cultivars by *Puccinia striiformis*: First report on disease severity and yield loss/Blatt-und Äenbefall sowie Ertragsverluste durch *Puccinia striiformis* in Triticale-Sorten. *Z. Für Pflanzenkrankh. Und Pflanzenschutz/J. Plant Dis. Prot.* **2004**, *111*, 461–464.
10. McIntosh, R.; Yamazaki, Y.; Devos, K.M.; Dubcovsky, J.; Rogers, W.J.; Appels, R. *Catalogue of Gene Symbols for Wheat*; Akadémiai Kiadó: Budapest, Hungary, 2003; Volume 49.
11. Adonina, I.G.; Orlovskaya, O.A.; Tereshchenko, O.Y.; Koren, L.V.; Khotyleva, L.V.; Shumny, V.K.; Salina, E.A. Development of commercially valuable traits in hexaploid triticale lines with *Aegilops* introgressions as dependent on the genome composition. *Russ. J. Genet.* **2011**, *47*, 453–461. [CrossRef]
12. Kwiatek, M.; Belter, J.; Majka, M.; Wiśniewska, H. Allocation of the S-genome chromosomes of *Aegilops variabilis* Eig. carrying powdery mildew resistance in triticale (×*Triticosecale* Wittmack). *Protoplasma* **2016**, *253*, 329–343. [CrossRef]
13. Kwiatek, M.; Majka, M.; Wiśniewska, H.; Apolinarska, B.; Belter, J. Effective transfer of chromosomes carrying leaf rust resistance genes from *Aegilops tauschii* Coss. into hexaploid triticale (X *Triticosecale* Witt.) using *Ae. tauschii* × *Secale cereale* amphiploid forms. *J. Appl. Genet.* **2015**, *56*, 163–168. [CrossRef]
14. Salmanowicz, B.P.; Langner, M.; Wiśniewska, H.; Apolinarska, B.; Kwiatek, M.; Błaszczyk, L. Molecular, physicochemical and rheological characteristics of introgressive *Triticale/Triticum monococcum* ssp. *monococcum lines with wheat 1D/1A chromosome substitution. Int. J. Mol. Sci.* **2013**, *14*, 15595–15614.
15. Kang, H.; Wang, H.; Huang, J.; Wang, Y.; Li, D.; Diao, C.; Zhu, W.; Tang, Y.; Wang, Y.; Fan, X.; et al. Divergent Development of Hexaploid Triticale by a Wheat–Rye–*Psathyrostachys huashanica* Trigeneric Hybrid Method. *PLoS ONE* **2016**, *11*, e0155667. [CrossRef]
16. Schneider, A.; Molnár, I.; Molnár-Láng, M. Utilisation of *Aegilops* (goatgrass) species to widen the genetic diversity of cultivated wheat. *Euphytica* **2008**, *163*, 1–19. [CrossRef]
17. Prażak, R.; Paczos-Grzęda, E. Gene transfer from *Aegilops kotschyi* Boiss. to *Triticum aestivum* L. *Adv. Agric. Sci. Probl.* **2010**, *555*, 613–620.
18. Rawat, N.; Neelam, K.; Tiwari, V.K.; Randhawa, G.S.; Friebe, B.; Gill, B.S.; Dhaliwal, H.S. Development and molecular characterization of wheat-*Aegilops kotschyi* addition and substitution lines with high grain protein, iron, and zinc. *Genome* **2011**, *54*, 943–953. [CrossRef]
19. Antonov, A.I.; Marais, G.F. Identification of leaf rust resistance genes in *Triticum* species for transfer to common wheat. *S. Afr. J. Plant Soil* **1996**, *13*, 55–60. [CrossRef]
20. Marais, G.F.; McCallum, B.; Snyman, J.E.; Pretorius, Z.A.; Marais, A.S. Leaf rust and stripe rust resistance genes *Lr54* and *Yr37* transferred to wheat from *Aegilops kotschyi*. *Plant Breed.* **2005**, *124*, 538–541. [CrossRef]
21. Heyns, I.; Pretorius, Z.A.; Marais, F. Derivation and Characterization of Recombinants of the *Lr54/Yr37* Translocation in Common Wheat. *Open Plant Sci. J.* **2011**, *5*, 1–8. [CrossRef]
22. Smit, C. Pyramiding of Novel Rust Resistance Genes in Wheat, Utilizing Marker Assisted Selection and Doubled Haploid Technology. Master's Thesis, Stellenbosch University, Stellenbosch, South Africa, 2013. RSA.

23. Qi, L.; Friebe, B.; Zhang, P.; Gill, B.S. Homoeologous recombination, chromosome engineering and crop improvement. *Chromosome Res.* **2007**, *15*, 3–19. [CrossRef]
24. Sears, E. Misdivision of univalents in common wheat. *Chromosoma* **1950**, *4*, 535–550. [CrossRef]
25. Sánchez-Monge, E. Two Types of Misdivision of the Centromere. *Nature* **1950**, *165*, 80–81. [CrossRef]
26. Friebe, B.; Zhang, P.; Linc, G.; Gill, B.S. Robertsonian translocations in wheat arise by centric misdivision of univalents at anaphase I and rejoining of broken centromeres during interkinesis of meiosis II. *Cytogenet. Genome Res.* **2005**, *109*, 293–297. [CrossRef]
27. Robertson, W.R.B. Chromosome studies. I. Taxonomic relationships shown in the chromosomes of tettigidae and acrididae: V-shaped chromosomes and their significance in *acrididae, locustidae,* and *gryllidae*: Chromosomes and variation. *J. Morph.* **1916**, *27*, 179–331. [CrossRef]
28. Lukaszewski, A.J.; Gustafson, J.P. Translocations andmodifications of chromosomes in triticale × wheat hybrids. *Theor. Appl. Genet.* **1983**, *64*, 239–248. [CrossRef]
29. Lukaszewski, A.J. Further manipulation by centric misdivision of the 1RS.1BL translocation in wheat. *Euphytica* **1997**, *94*, 257–261. [CrossRef]
30. Qi, L.L.; Pumphrey, M.O.; Friebe, B.; Zhang, P.; Qian, C.; Bowden, R.L.; Rouse, M.N.; Jin, Y.; Gill, B.S. A novel Robertsonian translocation event leads to transfer of a stem rust resistance gene (*Sr52*) effective against race *Ug99* from *Dasypyrum villosum* into bread wheat. *Theor. Appl. Genet.* **2011**, *123*, 159–167. [CrossRef]
31. Türkösi, E.; Darko, E.; Rakszegi, M.; Molnár, I.; Molnár-Láng, M.; Cseh, A. Development of a new 7BS.7HL winter wheat-winter barley Robertsonian translocation line conferring increased salt tolerance and (1,3;1,4)-β-D-glucan content. *PLoS ONE* **2018**, *13*, e0206248.
32. Lukaszewski, A.J.; Curtis, C.A. Transfer of the Glu-D1 gene fromchromosome 1D of bread wheat to chromosome 1R in hexaploidtriticale. *Plant Breed.* **1992**, *109*, 203–210. [CrossRef]
33. Lukaszewski, A.J. Manipulation of the 1RS.1BL translocation in wheat by induced homoeologous recombination. *Crop Sci.* **2000**, *40*, 216–225. [CrossRef]
34. Lukaszewski, A.J. Cytogenetically engineered rye chromosomes 1R to improve bread-making quality of hexaploid triticale. *Crop Sci.* **2006**, *46*, 2183–2194. [CrossRef]
35. Kwiatek, M.T.; Wiśniewska, H.; Ślusarkiewicz-Jarzina, A.; Majka, J.; Majka, M.; Belter, J.; Pudelska, H. Gametocidal Factor Transferred from *Aegilops geniculata* Roth Can Be Adapted for Large-Scale Chromosome Manipulations in Cereals. *Front. Plant Sci.* **2017**, *8*, 409. [CrossRef]
36. Heckmann, S.; Jankowska, M.; Schubert, V.; Kumke, K.; Ma, W.; Houben, A. Alternative meiotic chromatid segregation in the holocentric plant *Luzula elegans*. *Nat. Commun.* **2014**, *5*, 4979. [CrossRef]
37. Zwierzykowski, Z.; Zwierzykowska, E.; Taciak, M.; Jones, N.; Kosmala, A.; Krajewski, P. Chromosome pairing in allotetraploid hybrids of *Festuca pratensis* x *Lolium perenne* revealed by genomic in situ hybridization (GISH). *Chromosome Res.* **2008**, *16*, 575–585. [CrossRef]
38. Komuro, S.; Endo, R.; Shikata, K.; Kato, A. Genomic and chromosomal distribution patterns of various repeated DNA sequences in wheat revealed by a fluorescence in situ hybridization procedure. *Genome* **2013**, *56*, 131–137. [CrossRef] [PubMed]
39. Kwiatek, M.T.; Kurasiak-Popowska, D.; Mikołajczyk, S.; Niemann, J.; Tomkowiak, A.; Weigt, D.; Nawracała, J. Cytological markers used for identification and transfer of *Aegilops* spp. chromatin carrying valuable genes into cultivated forms of *Triticum*. *Comp. Cytogenet.* **2019**, *13*, 41–59. [CrossRef] [PubMed]
40. Ardalani, S.; Mirzaghaderi, G.; Badakhshan, H. A Robertsonian translocation from *Thinopyrum bessarabicum* into bread wheat confers high iron and zinc contents. *Plant Breed.* **2016**, *135*, 286–290. [CrossRef]
41. Tanaka, H.; Nabeuchi, C.; Kurogaki, M.; Garg, M.; Saito, M.; Ishikawa, G.; Nakamura, T.; Tsujimoto, H. A novel compensating wheat–*Thinopyrum elongatum* Robertsonian translocation line with a positive effect on flour quality. *Breed. Sci.* **2017**, *67*, 509–517. [CrossRef] [PubMed]
42. Danilova, T.V.; Friebe, B.; Gill, B.S.; Poland, J.; Jackson, E. Development of a complete set of wheat–barley group-7 Robertsonian translocation chromosomes conferring an increased content of β-glucan. *Theor. Appl. Genet.* **2018**, *131*, 377–388. [CrossRef] [PubMed]
43. Vega, J.M.; Feldman, M. Effect of the pairing gene *Ph1* on centromere misdivision in common wheat. *Genetics* **1998**, *148*, 1285–1294. [PubMed]
44. Steinitz-Sears, L.M. Somatic Instability of Telocentric Chromosomes in Wheat and the Nature of the Centromere. *Genetics* **1966**, *54*, 241–248.

45. Ghazali, S.; Mirzaghaderi, G.; Majdi, M. Production of a novel Robertsonian translocation from *Thinopyrum bessarabicum* into bread wheat. *Cytol. Genet.* **2015**, *49*, 378–381. [CrossRef]
46. Devos, K.M.; Atkinson, M.D.; Chinoy, C.N.; Francis, H.A.; Harcourt, R.L.; Koebner, R.M.D.; Liu, C.J.; Masojć, P.; Xie, D.X.; Gale, M.D. Chromosomal rearrangements in the rye genome relative to that of wheat. *Theor. Appl. Genet.* **1993**, *85*, 673–680. [CrossRef]
47. Friebe, B.; Hatchett, J.H.; Sears, R.G.; Gill, B.S. Transfer of Hessian fly resistance from 'Chaupon' rye to hexaploid wheat via a 2BS/2RL wheat-rye chromosome translocation. *Theor. Appl. Genet.* **1990**, *79*, 385–389. [CrossRef]
48. Rahmatov, M.; Rouse, M.N.; Nirmala, J.; Danilova, T.; Friebe, B.; Steffenson, B.J.; Johansson, E. A new 2DS•2RL Robertsonian translocation transfers stem rust resistance gene *Sr59* into wheat. *Theor. Appl. Genet.* **2016**, *129*, 1383–1392. [CrossRef] [PubMed]

Article

Physical Location of New Stripe Rust Resistance Gene(s) and PCR-Based Markers on Rye (*Secale cereale* L.) Chromosome 5 Using 5R Dissection Lines

Wei Xi [1,2,3,†], Zongxiang Tang [1,2,3,†], Jie Luo [1,2,3] and Shulan Fu [1,3,*]

1 College of Agronomy, Sichuan Agricultural University, Wenjiang, Chengdu 611130, China
2 Institute of Ecological Agriculture, Sichuan Agricultural University, Wenjiang, Chengdu 611130, China
3 Provincial Key Laboratory for Plant Genetics and Breeding, Wenjiang, Chengdu 611130, China
* Correspondence: fushulan@sicau.edu.cn; Tel.: +86-28-86290870
† These authors contributed equally to this work.

Received: 8 July 2019; Accepted: 26 August 2019; Published: 30 August 2019

Abstract: The rye (*Secale cereale* L.) 5R chromosome contains some elite genes that can be used to improve wheat cultivars. In this study, a set of $5R^{Ku}$ dissection lines was obtained, and 111 new PCR-based and $5R^{Ku}$-specific markers were developed using the specific length amplified fragment sequencing (SLAF-seq) method. The 111 markers were combined with the 52 $5R^{Ku}$-specific markers previously reported, and 65 *S. cereale* Lo7 scaffolds were physically mapped to six regions of the $5R^{Ku}$ chromosome using the $5R^{Ku}$ dissection lines. Additionally, the $5RL^{Ku}$ arm carried stripe rust resistance gene(s) and it was located to the region L2, the same region where 22 $5R^{Ku}$-specific markers and 11 *S. cereale* Lo7 scaffolds were mapped. The stripe rust resistance gene(s) located in the $5RL^{Ku}$ arm might be new one(s) because its source and location are different from the previously reported ones, and it enriches the resistance source of stripe rust for wheat breeding programs. The markers and the *S. cereale* Lo7 scaffolds that were mapped to the six regions of the $5R^{Ku}$ chromosome can facilitate the utilization of elite genes on the 5R chromosome in the improvement of wheat cultivars.

Keywords: wheat; rye; 5R dissection line; PCR-based markers; physical map; stripe rust

1. Introduction

Rye (*Secale cereale* L.) is a useful gene source supporting disease and insect resistance, stress resistance, and higher yield for wheat (*Triticum aestivum* L.) breeding programs [1–6]. In fact, rye 5R chromosome contains some elite genes that can be used to improve wheat cultivars. The wheat-rye 4BL/5RL translocation chromosome indicates that the 5RL arm carried loci that can increase copper efficiency [7]. The wheat-rye 5R(5A) and 5R(5D) substitution line exhibits some favorite traits, including big spike, multispikelet, high satiation seed, high protein content, and positive effects on somatic embryogenesis [8–10]. The rye 5RS arm carries the resistance gene to Russian wheat aphid (RWA) [11]. The $5R^aS$ arm that was derived from *S. africanum* Stapf. carries genes for stripe rust resistance, increasing spike length and reducing grain hardness [12]. The gene *Ddw1* that can reduce plant height in rye is located on the 5RL arm and its close linkage markers have been found [13]. It was also reported that the 5R chromosome harbors genes that determine its growth habit [14]. However, the elite genes on 5R chromosomes have not been successfully used in wheat cultivars because of linkage drag and the non-compensating and low recombination frequency of 5R with its wheat homologues [15,16]. Elite genes can be utilized effectively when segments of 5R chromosomes have been transferred into wheat backgrounds and been accurately identified. Rye chromosome-specific markers are conducive to identifying rye chromatin in wheat backgrounds and help in the effective application of rye elite genes in wheat breeding programs. So far, few 5R-specific markers have been developed. Although terminal deletions of 5RL (T5RS.5RL-del) were obtained, they were not used to physically map 5R-specific markers [17]. In this study, the $5R^{Ku}$ chromosome derived from rye Kustro (*S. cereale* L.) was proven to carry stripe rust resistance gene(s), and some new PCR-based and $5R^{Ku}$-specific markers were developed using specific length amplified fragment sequencing (SLAF-seq) technology. These new markers were mapped to six regions on $5R^{Ku}$ using $5R^{Ku}$ dissection lines in a wheat background. Additionally, the stripe rust resistance gene(s) was also physically located on a segment of $5R^{Ku}$.

2. Materials and Methods

2.1. Plant Materials

The octoploid triticale line MK was derived from common wheat (*T. aestivum* L.) Mianyang 11(MY11) × *S. cereale* L. Kustro. Some progeny was obtained by controlled backcrossing of MK with MY11 followed by strict self-seeding [18]. Seven wheat-rye monosomic addition lines, including $MA1R^{Ku}$, $MA2R^{Ku}$, $MA3R^{Ku}$, $MA4R^{Ku}$, $MA5R^{Ku}$, $MA6R^{Ku}$, and $MA7R^{Ku}$, were developed according to the methods described by Li et al. [18]. From the self-seeded progeny of the monosomic addition line $MA5R^{Ku}$, a $5R^{Ku}$(5B) disomic substitution line, $5RS^{Ku}$(5B) monotelosomic substitution line, $5RL^{Ku}$(5B) ditelosomic substitution line, $5RS^{Ku}$/5BL translocation line, and four $5R^{Ku}$ deletion lines were obtained. The progeny of $MA5R^{Ku}$ comprised a set of $5R^{Ku}$ dissection lines. In addition, rye Kustro (*S. cereale* L.), common wheat (*T. aestivum* L.) Chinese Spring (CS), and MY11 were also used in this study. The materials used in this study are available upon request to interested researchers.

2.2. Non-Denaturing Fluorescence in situ Hybridization (ND-FISH) Analysis

Oligonucleotide (oligo) probes, including Oligo-Ku [19], Oligo-pSc119.2-1, Oligo-pTa535-1 [20], Oligo-pSc200, and Oligo-pSc250 [21], were used for ND-FISH analysis of the materials used in this study. Oligo-pSc119.2-1 and Oligo-pTa535-1 can be used to identify individual wheat chromosomes [20]. Oligo-Ku, Oligo-pSc200, and Oligo-pSc250 can together be used to distinguish rye chromosomes from the wheat genome [19]. In this study, the combination of Oligo-pSc200 and Oligo-pSc250 was denoted as Oligo-pSc200 + 250. Oligo probes were synthesized by Tsingke Biological Technology Co. Ltd. (Beijing, China), and they were 5′-end-labeled with 6-carboxytetramethylrhodamine (TAMRA), 6-carboxyfluorescein (6-FAM), or Cyanine Dye 5 (Cy5). The chromosome spreads of the root tips were prepared following the methods described by Han et al. [22]. Images were made using an epifluorescence Olympus BX51 microscope (Olympus Corporation, Tokyo, Japan) equipped with a cooled charge-coupled device camera and with the HCIMAGE Live software (Hamamatsu Corporation, Sewickley, PA, USA). Images were processed using Adobe Photoshop CS 3.0.

2.3. Developing PCR-Based Markers

Genomic DNA of *S. cereale* L. Kustro has already been sequenced using specific length amplified fragment sequencing (SLAF-seq) technology (Biomarker, Beijing, China) [4]. In this study, the genomic DNA of wheat-rye monosomic addition line $MA5R^{Ku}$ was also sequenced using the SLAF-seq technique. Sixty base sequences of both ends of the sequences with sizes between 450 to 500 bp were obtained. The sequencing procedure was carried out according to the methods described by Duan et al. [4]. The primary $5R^{Ku}$ specific pair-end reads were obtained according to the methods described by Duan et al. [4]. Primers were designed according to 542 randomly selected $5R^{Ku}$ specific pair-end reads using the software Primer 3 (version 4.1.0) [23], and the optimal melting temperature and size values were set to 60 °C and 20 bases, respectively. The 542 pair-end reads were deposited in the GenBank Database (GenBank accession numbers: MN325158-MN325699). A total of 542 primer pairs were designed.

2.4. PCR Assay, 5R^{Ku}-Specific Markers Testing and Physical Location

The PCR amplifications and electrophoresis were carried out according to the procedure described by Duan et al. [4]. The markers that could produce bands from both rye Kustro and MA5R^{Ku} but not from CS, MY11, MA1R^{Ku}, MA2R^{Ku}, MA3R^{Ku}, MA4R^{Ku}, MA6R^{Ku}, and MA7R^{Ku} were regarded as 5R^{Ku}-specific markers. For each of the primer pairs, PCR reactions were repeated three times. These markers combined with the 52 5R^{Ku}-specific markers developed by Qiu et al. [24] were mapped to specific regions of the 5R^{Ku} chromosome using the 5R^{Ku} dissection lines. Again, PCR reactions of each 5R^{Ku}-specific marker were repeated three times.

2.5. Similarity Searches of the Pair-End Reads Against S. cereale L. Lo7 Scaffolds

The original pair-end reads for designing the 5R^{Ku}-specific markers were used for Nucleotide BLAST searching against the sequences in the *S. cereale* L. Lo7 scaffolds database in GrainGenes [25]. The *S. cereale* L. Lo7 scaffolds were reported by Bauer et al. [26].

2.6. Stripe Rust Resistance Testing

The resistance of 5R^{Ku} dissection lines and the parental wheat MY11 to stripe rust was evaluated under field conditions. Plants were grown in Qionglai, Sichuan, China. The materials were inoculated twice with the mixed stripe rust prevalent isolates CYR32, CYR33, and CYR34 in China at approximately 6 weeks and 9 weeks after sowing. Infection types (IT) were scored according to a 0–9 numerical scale as described by Wan et al. [27] in the adult stage.

3. Results

3.1. Obtaining 5R^{Ku} Dissection Lines

From the self-seeded progeny of MA5R^{Ku}, lines 142-30, 142-77, 143-9, 143-61, 237-2, 403-50, 444-12, and 449-4 were obtained (Figures 1–3; Table 1). Probe Oligo-pSc119.2-1 combined with probe Oligo-pTa535-1 can identify individual wheat chromosomes (Figures 1–3). The combination of probes, Oligo-Ku, Oligo-pSc200, Oligo-pSc250, and Oligo-pSc119.2-1, can be used to distinguish rye 5R^{Ku} chromosomes from the wheat genome and detect their structural variations (Figures 1–3). For the sake of description, the combination of Oligo-pSc200 and Oligo-pSc250 was denoted as Oligo-pSc200 + 250. Wheat 5B chromosomes disappeared from the five lines 140-30, 142-77, 143-9, 143-61, and 403-50 (Figures 1 and 2A–D). A broken 5R^{Ku} chromosome existed in line 142-30, and the breakpoint located between the interstitial Oligo-pSc119.2-1 and the Oligo-pSc200 + 250 signaled sites of the 5RL^{Ku} arm (Figure 1A,B). Line 142-77 contained a 5RS^{Ku} arm, therefore, it was a 5RS^{Ku}(5B) monotelosomic substitution line (Figure 1C,D). Two broken 5R^{Ku} chromosomes existed in line 143-9, and the breakpoints were located between the Oligo-pSc200 + 250 signal sites and the telomeres of 5RL^{Ku} arms (Figure 1E,F). Lines 143-61, 403-50, and 237-2 were 5R^{Ku}(5B) disomic substitution, 5RL^{Ku}(5B) ditelosomic substitution, and 5RS^{Ku}/5BL translocation lines, respectively (Figure 2). Additionally, a 7B chromosome in line 142-30 had no short arm (Figure 1A) and a wheat 3D chromosome disappeared from line 143-9 (Figure 1F).

Line 444-12 contained two broken $5R^{Ku}$ chromosomes, and the breakpoints were located between the centromeres and the interstitial Oligo-pSc119.2-1 signal sites (Figure 3A,B). In line 449-4, another broken $5R^{Ku}$ chromosome was observed, and the breakpoint occurred on the $5RS^{Ku}$ arm (Figure 3C,D). A wheat 3D chromosome also disappeared from line 449-4 (Figure 3C,D). The chromosome compositions of the eight lines are listed in Table 1. According to the breakpoints, the $5R^{Ku}$ chromosome could be divided into six regions (Figure 4). The broken $5R^{Ku}$ chromosomes in lines 142-30, 143-9, 444-12, and 449-4 were named $Del5R^{Ku}L^{3-4}$, $Del5R^{Ku}L^{4}$, $Del5R^{Ku}L^{2-4}$, and $Del5R^{Ku}S^{1}$, respectively (Figure 4). Therefore, 142-30, 142-77, 143-9, 143-61, 237-2, 403-50, 444-12, and 449-4 comprised a set of $5R^{Ku}$ dissection lines.

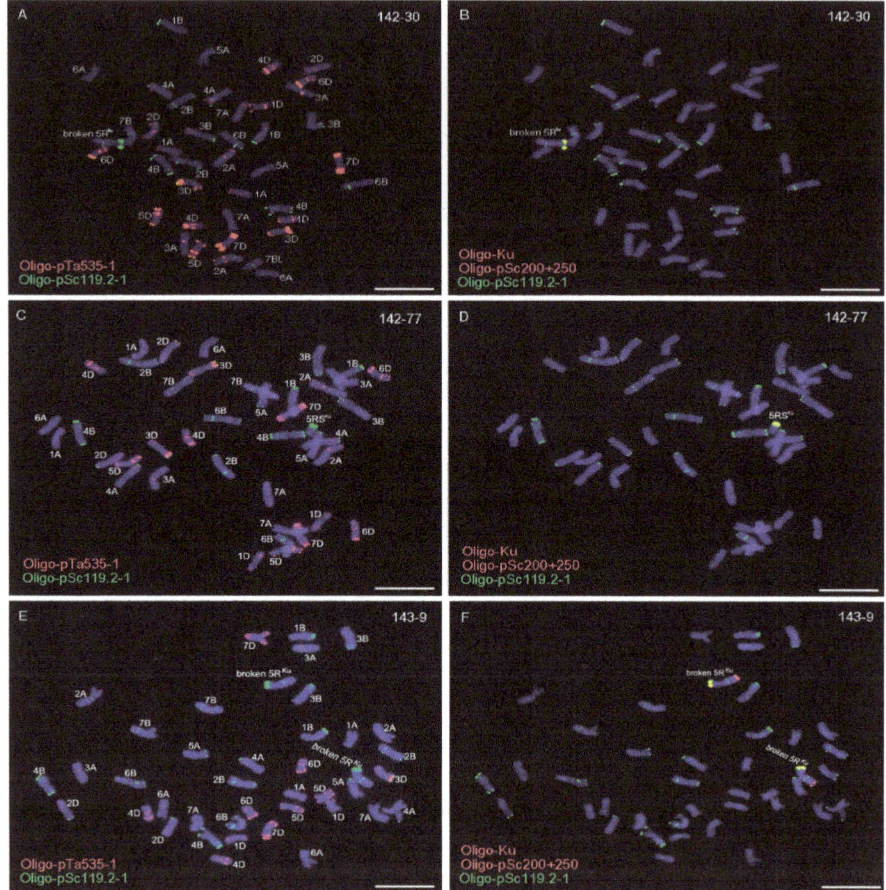

Figure 1. ND-FISH analysis of root tip metaphase chromosomes of lines 142-30, 142-77, and 143-9 using Oligo-pTa535-1, Oligo-pSc119.2-1, Oligo-Ku, and Oligo-pSc200 + 250 as probes. (**A**) and (**B**) are the same cell, (**C**) and (**D**) are the same cell, and (**E**) and (**F**) are the same cell. Chromosomes were counterstained with 4′,6′-diamidino-2-phenylindole (DAPI) (blue). Scale bar is 10 μm.

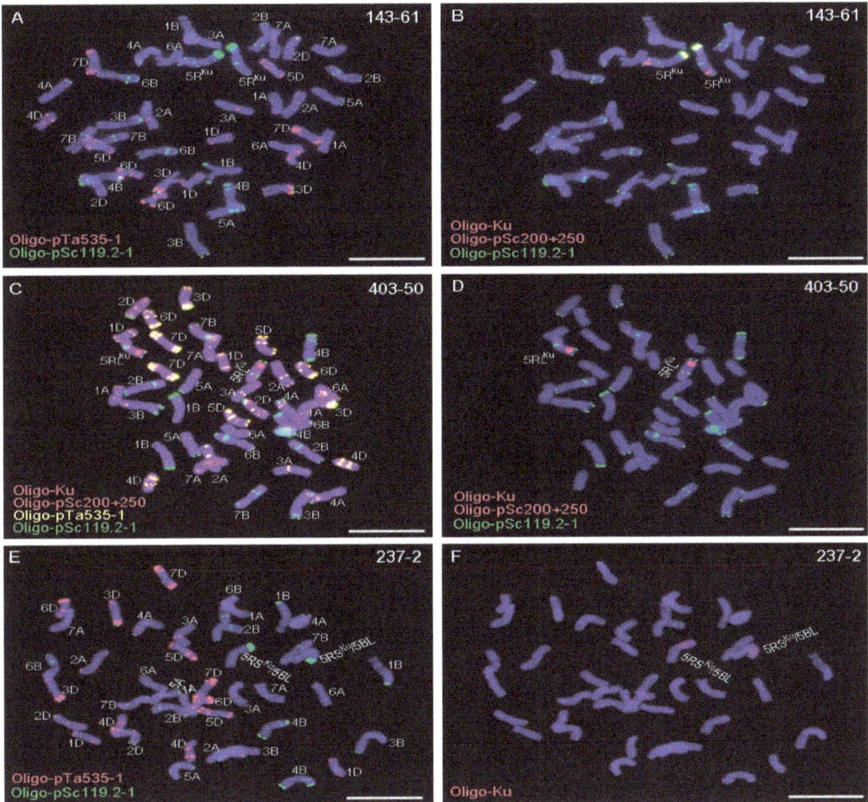

Figure 2. ND-FISH analysis of root tip metaphase chromosomes of lines 143-61, 403-50, and 237-2 using Oligo-pTa535-1, Oligo-pSc119.2-1, Oligo-Ku, and Oligo-pSc200 + 250 as probes. (**A**) and (**B**) are the same cell, (**C**) and (**D**) are the same cell, and (**E**) and (**F**) are the same cell. Chromosomes were counterstained with DAPI (blue). Scale bar is 10 μm.

Table 1. Information of the materials used in this study.

Name	Chromosome Constitution
142-30	one broken $5R^{Ku}$, 39 intact wheat chromosomes and one 7BL arm
142-77	one $5RS^{Ku}$ arm, 40 intact wheat chromosomes
143-9	two broken $5R^{Ku}$ chromosomes, 39 intact wheat chromosomes
143-61	two intact $5R^{Ku}$ chromosomes, 40 intact wheat chromosomes
237-2	two $5RS^{Ku}$/5BL translocation chromosomes, 40 intact wheat chromosomes
403-50	two $5RL^{Ku}$ arms, 40 intact wheat chromosomes
444-12	two broken $5R^{Ku}$ chromosomes, 42 intact wheat chromosomes
449-4	one broken $5R^{Ku}$ chromosome, 41 intact wheat chromosomes

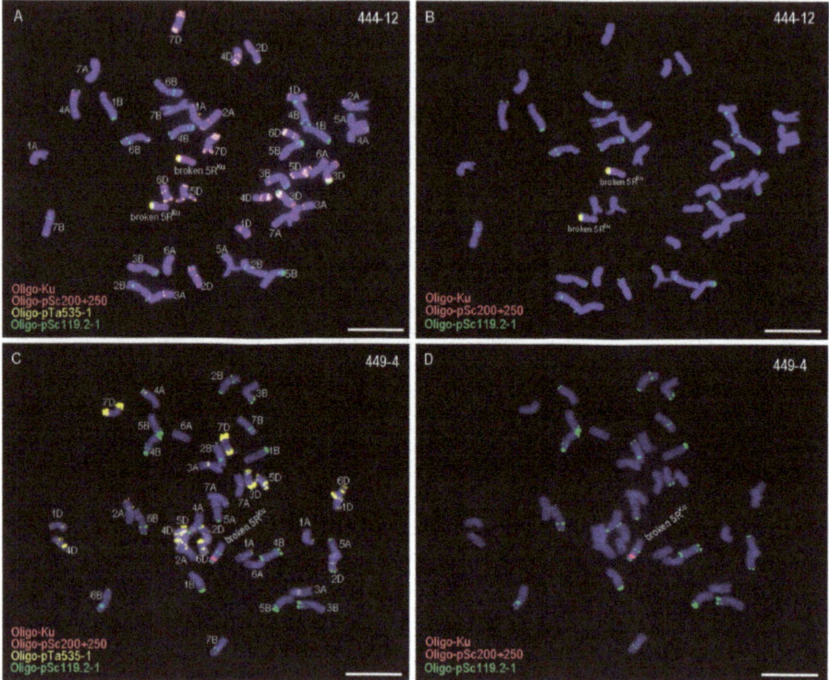

Figure 3. ND-FISH analysis of root tip metaphase chromosomes of lines 444-12 and 449-4 using Oligo-pTa535-1, Oligo-pSc119.2-1, Oligo-Ku, and Oligo-pSc200 + 250 as probes. (**A**) and (**B**) are the same cell, and (**C**) and (**D**) are the same cell. Chromosomes were counterstained with DAPI (blue). Scale bar is 10 μm.

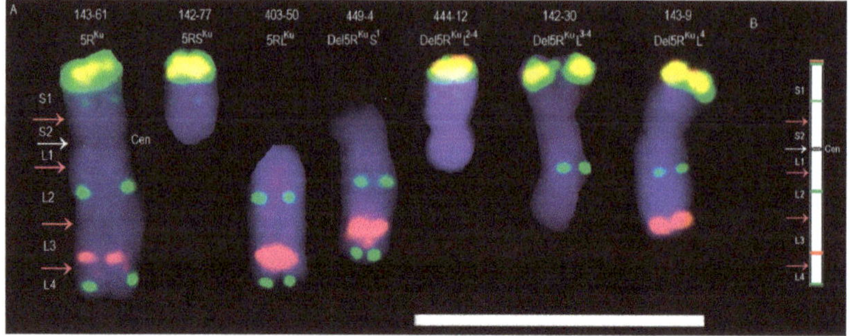

Figure 4. Cut-pasted $5R^{Ku}$ dissections and the schematic diagram of $5R^{Ku}$ chromosome. (**A**) Cut-pasted $5R^{Ku}$, $5RS^{Ku}$, $5RL^{Ku}$, and the four kinds of broken $5R^{Ku}$. 5RKu chromosome was divided into six regions. (**B**) The schematic diagram of $5R^{Ku}$ chromosome. Red arrows indicate broken points. White arrow and "Cen" indicate centromere. Scale bar is 10 μm.

3.2. Developing $5R^{Ku}$-Specific Markers and Physical Mapping

Rye Kustro, CS, MY11, and the seven wheat-rye monosomic addition lines were used to develop $5R^{Ku}$-specific markers. In total, 11 of the 542 primer pairs amplified specific bands from rye Kustro and $MA5R^{Ku}$, but not from CS, MY11, $MA1R^{Ku}$, $MA2R^{Ku}$, $MA3R^{Ku}$, $MA4R^{Ku}$, $MA6R^{Ku}$, and $MA7R^{Ku}$ (Figure 5).

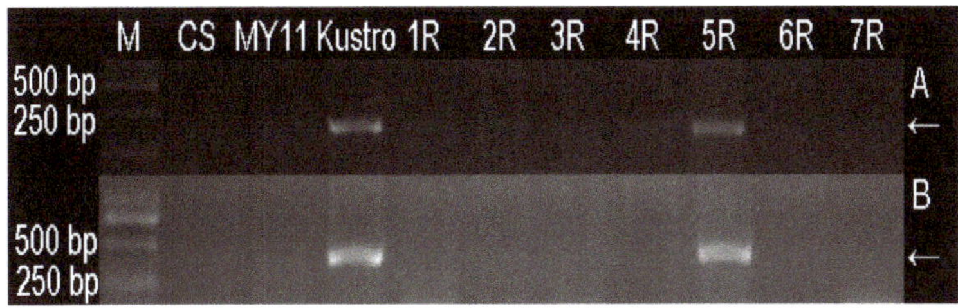

Figure 5. Developing $5R^{Ku}$-specific markers. (**A**) Products amplified by marker Ku5R-58. (**B**) Products amplified by marker Ku5R-516. M: DNA marker; CS: Chinese Spring; MY11: Mianyang 11; Kustro: rye kustro; 1R-7R: $MA1R^{Ku}$-$MA7R^{Ku}$. Arrows indicate the target bands.

Therefore, the 111 primer pairs were regarded as $5R^{Ku}$-specific markers and the information of these markers is listed in Table S1. Subsequently, the 111 markers and the 52 $5R^{Ku}$-specific markers developed by Qiu et al. [24] were physically mapped to six regions of the $5R^{Ku}$ chromosome using lines 142-30, 142-77, 143-9, 143-61, 237-2, 403-50, 444-12, and 449-4 (Figures 6 and 7; Table S2).

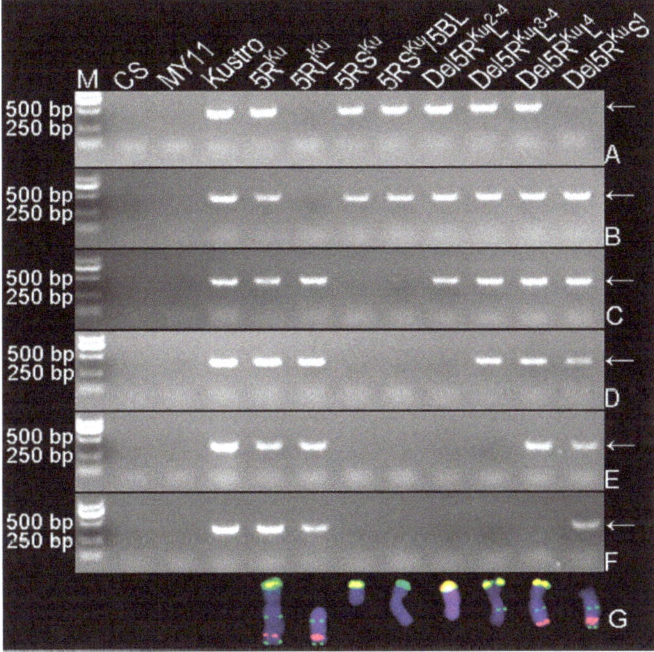

Figure 6. Physically localized $5R^{Ku}$-specific markers using $5R^{Ku}$ dissection lines. (**A**) Products amplified by primer pair Ku5R-290 representing the markers mapped to region S1. (**B**) Products amplified by primer pair Ku5R-271 representing the markers mapped to region S2. (**C**) Products amplified by primer pair Ku5R-120 representing the markers mapped to region L1. (**D**) Products amplified by primer pair Ku5R-342 representing the markers mapped to region L2. (**E**) Products amplified by primer pair Ku5R-48 representing the markers mapped to region L3. (**F**) Products amplified by primer pair Ku5R-9 representing the markers mapped to region L4. (**G**) Cut-pasted $5R^{Ku}$ chromosome, $5RS^{Ku}$ arm, $5RL^{Ku}$ arm, $5RS^{Ku}$/5BL translocation chromosome, and the four kinds of broken $5R^{Ku}$ chromosomes corresponding to their own amplified products in each electrophoresis lane. Arrows indicate the target bands amplified by each marker.

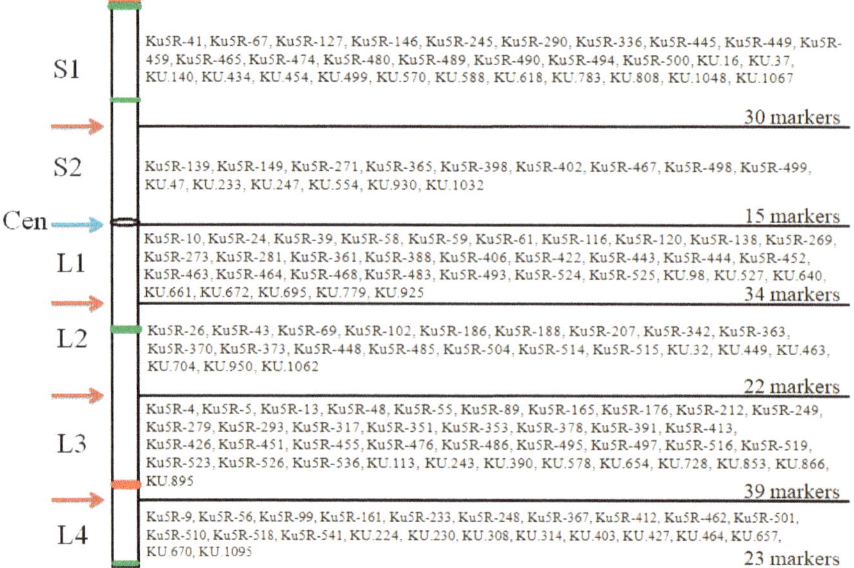

Figure 7. The schematic diagram for the physical map of $5R^{Ku}$-specific markers. These markers were mapped to six regions on $5R^{Ku}$ chromosome according to their amplicons in $5R^{Ku}$ dissection lines. The six regions are divided by five dark lines and named S1, S2, L1, L2, L3, and L4, respectively. The red arrows corresponding to each dark line indicate the breakpoints on the $5R^{Ku}$ chromosome. The blue arrow and "Cen" indicate the centromere. In each region, the names of the markers are listed on the right. In the schematic diagram, green bands represent FISH signals of Oligo-pSc119.2-1 and the red bands represent the FISH signals of Oligo-pSc200 + 250. The markers named "KU + number" were developed by Qiu et al. [24], and those named "Ku5R + number" were developed in this study.

Thirty markers amplified their target bands from lines 143-16 ($5R^{Ku}$), 142-77 ($5RS^{Ku}$), 237-2 ($5RS^{Ku}$/5BL), 444-12 ($Del5R^{Ku}L^{2-4}$), 142-30 ($Del5R^{Ku}L^{3-4}$), and 143-9 ($Del5R^{Ku}L^{4}$) but not from lines 403-50 ($5RL^{Ku}$) and 449-4 ($Del5R^{Ku}S^{1}$) (Figure 6A). So, the 30 markers were mapped to region S1 of the $5R^{Ku}$ chromosome (Figure 7). The target products of 15 markers only disappeared from line 403-50 ($5RL^{Ku}$); therefore, they were mapped to region S2 of the $5R^{Ku}$ chromosome (Figures 6B and 7). The target products of 34 markers only disappeared from lines 142-77 ($5RS^{Ku}$) and 237-2 ($5RS^{Ku}$/5BL), and they were mapped to region L1 of the $5R^{Ku}$ chromosome (Figures 6C and 7). The target products of 22 markers existed in lines 143-16 ($5R^{Ku}$), 403-50 ($5RL^{Ku}$), 142-30 ($Del5R^{Ku}L^{3-4}$), 143-9 ($Del5R^{Ku}L^{4}$), and 449-4 ($Del5R^{Ku}S^{1}$) but not in lines 142-77 ($5RS^{Ku}$), 237-2 ($5RS^{Ku}$/5BL), and 444-12 ($Del5R^{Ku}L^{2-4}$) (Figure 6D). Therefore, the 22 markers were mapped to region L2 of $5R^{Ku}$ chromosome (Figure 7). Thirty-nine markers did not amplify their target products from lines 142-77 ($5RS^{Ku}$), 237-2 ($5RS^{Ku}$/5BL), 444-12 ($Del5R^{Ku}L^{2-4}$), and 142-30 ($Del5R^{Ku}L^{3-4}$), and 23 markers did not amplify their target products from lines 142-77 ($5RS^{Ku}$), 237-2 ($5RS^{Ku}$/5BL), 444-12 ($Del5R^{Ku}L^{2-4}$), 142-30 ($Del5R^{Ku}L^{3-4}$), and 143-9 ($Del5R^{Ku}L^{4}$) (Figure 6E,F). Therefore, the 39 and 23 markers were mapped to regions L3 and L4 of $5R^{Ku}$ chromosome, respectively (Figure 7).

3.3. Similarity between S. cereale Lo7 Scaffolds and Pair-End Reads Used for $5R^{Ku}$-Specific Primers Design

The original pair-end reads that were used to design $5R^{Ku}$-specific primer pairs were deposited in the GenBank Database (GenBank accession numbers: MN325158-MN325268) (Table S1). The corresponding pair-end reads to $5R^{Ku}$-specific markers were used for nucleotide BLAST search against the *S. cereale* Lo7 scaffolds database using the blastn tool in GrainGenes [25] The *S. cereale* Lo7 scaffolds served to validate the chromosomal localization of the majority of the $5R^{Ku}$-specific markers. The results were listed in Table S1. The 111 pair-end reads had 83% to 100% similarity with some of the *S. cereale* Lo7 scaffolds (Table S1). The left and the right 1 to 60 bp nucleotide sequences of 13 pair-end reads, respectively, hit two Lo7 scaffolds that were derived from different rye chromosomes (Table S1). Twenty-two pair-end reads only hit the Lo7 scaffolds that derived from the 0R chromosome (Table S1). Four pair-end reads respectively hit the five Lo7 scaffolds that were derived from 1R, 3R, or 4R chromosomes (Table S1). The left and the right 1 to 60 bp nucleotide sequences of seven pair-end reads hit two different Lo7 scaffolds that were derived from 5R chromosome (Table S1). Sixty-five pair-end reads hit 65 single Lo7 scaffolds that were derived from the 5R chromosome, respectively (Table S1). Although there were 72 5R-derived Lo7 scaffolds matching the pair-end reads, only 65 Lo7 scaffolds were mapped to six regions of the $5R^{Ku}$ chromosome because they hit single scaffolds (Figure 8).

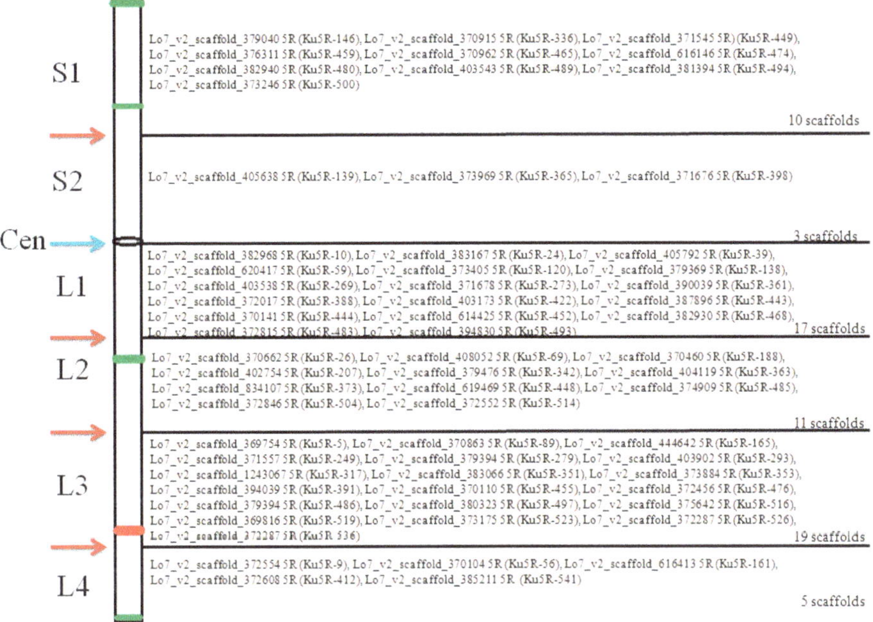

Figure 8. The schematic diagram for the physical map of *S. cereale* Lo7 scaffolds that were derived from the 5R chromosome. The names of the scaffolds are listed on the right and their corresponding markers are listed in brackets. The means of the other symbols are the same as the ones in Figure 7.

3.4. Location of Stripe Rust Gene(s) on $5R^{Ku}$ Chromosome

The resistance of lines 142-30, 142-77, 143-9, 143-61, 237-2, 403-50, 444-12, and 449-4 and parental wheat MY11 to stripe rust was tested. According to the standard scale of infection types that was described by Wan et al. [27], the infection types of lines 143-16 ($5R^{Ku}$), 403-50 ($5RL^{Ku}$), 142-30 ($Del5R^{Ku}L^{3-4}$), 143-9 ($Del5R^{Ku}L^{4}$), and 449-4 ($Del5R^{Ku}S^{1}$) were 2 to 3 (Figure 9). The infection types of MY11 and the lines 142-77 ($5RS^{Ku}$), 237-2 ($5RS^{Ku}$/5BL), and 444-12 ($Del5R^{Ku}L^{2-4}$) were 8 to 9 (Figure 9). These results indicated that the $5R^{Ku}$ chromosome carried stripe rust resistance gene(s) and it was located on the $5RL^{Ku}$ arm. Further, the resistance gene(s) was mapped to regions L2-L4 because line 444-12 ($Del5R^{Ku}L^{2-4}$) was highly susceptible to stripe rust (Figure 9). At last, the stripe rust resistance gene(s) could be mapped to region L2 of the $5R^{Ku}$ chromosome because lines 142-30 ($Del5R^{Ku}L^{3-4}$) and 143-9 ($Del5R^{Ku}L^{4}$) exhibited resistance to stripe rust (Figure 9).

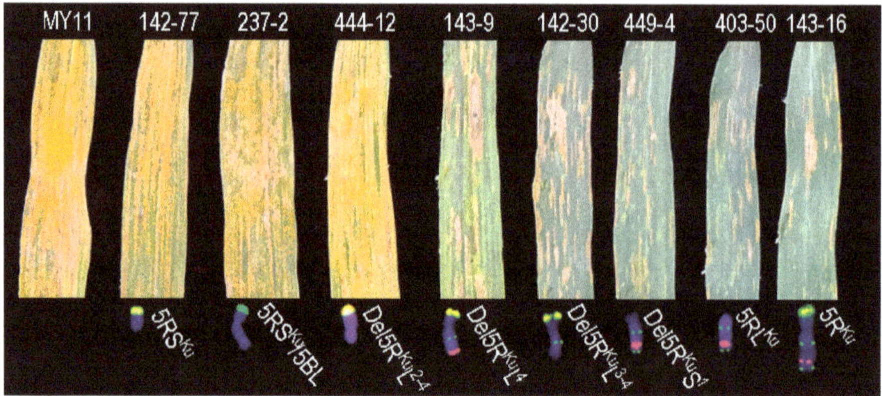

Figure 9. Stripe rust resistance testing. MY11, and lines 142-77, 237-2, and 444-12 are highly susceptible to stripe rust. Lines 143-9, 142-30, 449-4, 403-50, and 143-16 display resistance to stripe rust.

4. Discussion

4.1. 5R-Specific Markers

In our previous studies, some PCR-based and rye chromosome-specific markers have been developed using SLAF-seq technology [4,18,24]. Some of these markers have already been used to identify rye chromosomal segments in wheat backgrounds because they have easy application [5,6,28]. It has already been reported that rye 5R chromosomes carry some elite genes for wheat cultivar improvement [7–12]. Rye 5R-specific markers are contributive to identifying 5R chromatin in wheat backgrounds and help in the effective application of 5R elite genes in wheat breeding programs. So far, some PCR-based and 5R-specific markers have been reported. Two *S. cereale* inter-microsatellite (SCIM) markers were found to be 5R-specific [29]. Tomita et al. [30] reported a 5R-specific marker that was derived from the *Revolver* transposon-like sequence. A 5RS-specific and five 5RL-specific markers derived from expressed sequence tags (ESTs) were developed [31]. Sixteen PLUG markers that were developed using the PCR-based landmark unique gene system were proven to be 5R-specific [32]. Twenty-one $5R^{a}$-specific molecular markers were developed [12]. Qiu et al. [24] developed 19 $5RS^{Ku}$-specific and 33 $5RL^{Ku}$-specific markers using SLAF-seq technology. These markers mentioned above are easier to perform because they are PCR-based markers. In addition to distinguishing the 5R chromosome in wheat backgrounds, these markers can be used to construct a map of the 5R chromosome. However, more PCR-based and 5R-specific markers are needed. The 111 markers developed in this study have enriched the 5R-specific markers. In this study, a total of 163 $5R^{Ku}$-specific markers were mapped to six regions of the $5R^{Ku}$ chromosome, and this is beneficial for the identification of introgressed 5R small segments in wheat backgrounds. Additionally, the corresponding *S. cereale* Lo7 scaffolds [26] to 65 $5R^{Ku}$-specific markers were found and they were also mapped to the six regions of the $5R^{Ku}$ chromosome. This might be useful for building the physical map and the high-density genetic map of chromosome 5R. Furthermore, the *S. cereale* Lo7 scaffolds [26] mapped to the region L2 might help to further dissect the stripe resistance gene(s). Twenty-two of the 111 pair-end reads matched some scaffolds that have not yet been assigned to any rye chromosome (0R) and these pair-end reads contributed to mapping a reasonable number of *S. cereale* Lo7 scaffolds to the different regions of the 5R chromosome. The case that some $5R^{Ku}$-specific pair-end reads hit some Lo7 scaffolds derived from 1R, 3R, or 4R chromosomes might be the result from duplications, members of gene families, etc.

In addition, Silkova et al. reported some 5R deletion lines. However, their breakpoints were unclear, and these deletion lines were not used to map 5R-specific markers [17]. In this study, the signal patterns of probes Oligo-Ku, Oligo-pSc200, Oligo-pSc250, and Oligo-pSc119.2-1 on the rye chromosomes are the same as the ones of the $5R^{Ku}$ chromosomes reported by Li et al. [18]. The rye chromosomes in this study were determined to be chromosome 5R, although the control was not used. Therefore, a new set of 5R dissection lines was developed in this study and the breakpoints on these broken 5R chromosomes were tentatively determined.

4.2. Stripe Rust Resistance Gene(s) on the 5RL Arm Might be New One

It has been reported that 1R chromosomes that were derived from various rye sources carried stripe rust resistance gene(s) [33–37]. Additionally, the 2RL and 5RS arms of *S. africanum* Stapf. [12,38], the 4R chromosome of rye cultivar German White [6], and the 6R chromosome of *S. cereanum* cv. Kriszta [39] also carry stripe rust resistance gene(s). In this study, the $5RL^{Ku}$ arm that was derived from rye Kustro also carried stripe rust resistance gene(s) and the resistance gene(s) was mapped to region L2 of the $5R^{Ku}$ chromosome. Therefore, the stripe rust resistance gene(s) might be new one(s) because of its different source and location from the previously reported ones. The stripe rust resistance gene(s) on the $5RL^{Ku}$ arm reported in this study enriches the resistance source of stripe rust for wheat breeding programs. Additionally, 22 markers and 11 *S. cereale* Lo7 scaffolds were located to the L2 region of the $5R^{Ku}$ chromosome, the same region where the stripe rust resistance gene(s) was mapped, and these markers and scaffolds can facilitate the utilization of the resistance gene(s).

4.3. Variations of Wheat Chromosomes

Compared with the standard signal patterns of probes Oligo-pSc119.2-1 and Oligo-pTa535-1 on the chromosomes of common wheat MY11 [20], it can be noted that the variations of wheat chromosomes occurred in lines 142-30, 143-9, and 449-4. This indicated that wheat-rye 5R addition lines or 5R(5B) substitution lines could cause changes of the wheat genome. However, only alterations of the 7B and 3D chromosomes in the three lines were observed. Therefore, the rules and mechanisms of the alterations of the wheat genome in 5R addition and 5R(5B) substitution lines are not clear, and more attention should be paid to this issue in future research.

5. Conclusions

In the present study, a new set of $5R^{Ku}$ dissection lines was identified, and 111 new PCR-based and $5R^{Ku}$-specific markers were developed. The 111 markers were combined with the 52 $5R^{Ku}$-specific markers reported previously, and 65 *S. cereale* Lo7 scaffolds were physically mapped to six regions of the $5R^{Ku}$ chromosome. Additionally, the $5RL^{Ku}$ arm carried stripe rust resistance gene(s) and it was mapped to region L2 of the $5R^{Ku}$ chromosome, the same region where 22 $5R^{Ku}$-specific markers and 11 *S. cereale* Lo7 scaffolds were mapped. The stripe rust resistance gene(s) located in the $5RL^{Ku}$ arm enriches the resistance source of stripe rust for wheat breeding programs, and the markers and the 11 *S. cereale* Lo7 scaffolds that were mapped to the L2 region of the $5R^{Ku}$ chromosome facilitates the utilization of the stripe rust resistance gene(s) in the improvement of wheat cultivars.

Supplementary Materials: The following are available online at http://www.mdpi.com/2073-4395/9/9/498/s1, Table S1: The information of $5R^{Ku}$-specific markers. Table S2: Location of 5RKu-specific markers on six regions of $5R^{Ku}$ chromosome using $5R^{Ku}$ dissection lines.

Author Contributions: Conceived and designed the study, S.F. and Z.T.; performed the experiments, W.X. and J.L.; data analysis, W.X.; writing-original draft preparation, Z.T.; writing-review and editing, S.F.

Funding: This manuscript is provided by the National Key Research and Development Program of China (No. 2016YFD0102001) and National Natural Science Foundation of China (No. 31770373).

Acknowledgments: We are thankful to Yanling Duan, Ruiying Liao and Shuyao Tang at the College of Agronomy, Sichuan Agricultural University for their helpful work on nucleotide blast searching and data submission to NCBI.

Conflicts of Interest: The authors declare no conflict of interest.

References

1. Dundas, I.S.; Frappell, D.E.; Crack, D.M.; Fisher, J.M. Deletion mapping of a nematode resistance gene on rye chromosome 6R in wheat. *Crop Sci.* **2001**, *41*, 1771–1778. [CrossRef]
2. Yokosho, K.; Yamaji, N.; Ma, J.F. Isolation and characterisation of two MATE genes in rye. *Funct. Plant Biol.* **2010**, *37*, 296–303. [CrossRef]
3. Howell, T.; Hale, I.; Jankuloski, L.; Bonafede, M.; Gilbert, M.; Dubcovsky, J. Mapping a region within the 1RS.1BL translocation in common wheat affecting grain yield and canopy water status. *Theor. Appl. Genet.* **2014**, *127*, 2695–2709. [CrossRef] [PubMed]
4. Duan, Q.; Wang, Y.Y.; Qiu, L.; Ren, T.H.; Li, Z.; Fu, S.L.; Tang, Z.X. Physical location of new PCR-based markers and powdery mildew resistance gene(s) on rye (*Secale cereale* L.) chromosome 4 using 4R dissection lines. *Front. Plant Sci.* **2017**, *8*, 1716. [CrossRef] [PubMed]
5. Du, H.; Tang, Z.; Duan, Q.; Tang, S.; Fu, S. Using the $6RL^{Ku}$ minichromosome of rye (*Secale cereale* L.) to create wheat-rye $6D/6RL^{Ku}$ small segment translocation lines with powdery mildew resistance. *Int. J. Mol. Sci.* **2018**, *19*, 3933. [CrossRef] [PubMed]
6. An, D.; Ma, P.; Zheng, Q.; Fu, S.; Li, L.; Han, F.; Han, G.; Wang, J.; Xu, Y.; Jin, Y.; et al. Development and molecular cytogenetic identification of a new wheat-rye 4R chromosome disomic addition line with resistances to powdery mildew, stripe rust and sharp eyespot. *Theor. Appl. Genet.* **2019**, *132*, 257–272. [CrossRef] [PubMed]
7. Schlegel, R.; Werner, T.; Hülgenhof, E. Confirmation of a 4BL/5RL wheat-rye chromosome translocation line in the wheat cultivar 'Viking' showing high copper efficiency. *Plant Breed.* **1991**, *107*, 226–234. [CrossRef]
8. Sibikeev, S.N.; Sibikeeva, Y.E.; Krupnov, V.A. Transmission of 5R chromosomes via gametes and its influence on spring bread wheat somatic embryoidogenesis in vitro. *Russ. J. Genet.* **2005**, *41*, 1650–1655. [CrossRef]
9. Li, J.L.; Wang, X.P.; Zhong, L.; Xu, X.L. Study on homoeologous chromosome pairing and translocation induced by 5A/5R × 6A/6R wheat-rye substitution lines. *Acta Genet. Sin.* **2006**, *33*, 244–250. [CrossRef]
10. Chumanova, E.V.; Efremova, T.T.; Trubacheeva, N.V.; Arbuzova, V.S.; Rosseeva, L.P. Chromosome composition of wheat-rye lines and the influence of rye chromosomes on disease resistance and agronomic traits. *Russ. J. Genet.* **2014**, *50*, 1169–1178. [CrossRef]
11. Andersson, S.C.; Johansson, E.; Baum, M.; Rihawi, F.; Bouhssini, M.E.L. New resistance sources to Russian wheat aphid (*Diuraphis noxia*) in Swedish wheat substitution and translocation lines with rye (*Secale cereale*) and *Leymus mollis*. *Czech J. Genet. Plant Breed.* **2015**, *51*, 162–165. [CrossRef]
12. Li, G.; Gao, D.; La, S.; Wang, H.; Li, J.; He, W.; Yang, E.; Yang, Z. Characterization of wheat-*Secale africanum* chromosome $5R^a$ derivatives carrying *Secale specific* genes for grain hardness. *Planta* **2016**, *243*, 1203–1212. [CrossRef] [PubMed]
13. Braun, E.M.; Tsvetkova, N.; Rotter, B.; Siekmann, D.; Schwefel, K.; Krezdorn, N.; Plieske, J.; Winter, P.; Melz, G.; Voylokov, A.V.; et al. Gene expression profiling and fine mapping identifies a gibberellin 2-Oxidase gene co-segregating with the dominant dwarfing gene *Ddw1* in rye (*Secale cereale* L.). *Front. Plant Sci.* **2019**, *10*, 857. [CrossRef] [PubMed]
14. Konovalov, A.A.; Moiseeva, E.A.; Goncharov, N.P.; Kondratenko, E.Y. The order of the *bs*, *Skdh*, and *Aadh1* genes in chromosome 5R of rye *secale cereal* L. *Russ. J. Genet.* **2010**, *46*, 666–671. [CrossRef]
15. Lukaszewski, A.J.; Porter, D.R.; Baker, C.A.; Rybka, K.; Lapinski, B. Attempts to transfer Russion wheat aphid resistance from a rye chromosome in Russian triticales to wheat. *Crop Sci.* **2001**, *41*, 1743–1749. [CrossRef]
16. Lukaszewski, A.J. *Alien Introgression in Wheat*, 1st ed.; Molnár-Láng, M., Ceoloni, C., Doležel, J., Eds.; Springer: Cham, Switzerland, 2015; Chapter 7; pp. 163–189.
17. Silkova, O.G.; Leonova, I.N.; Krasilova, N.M.; Dubovets, N.I. Preferential elimination of chromosome 5R of rye in the progeny of 5R5D dimonosomics. *Russ. J. Genet.* **2011**, *47*, 942–950. [CrossRef]
18. Li, M.; Tang, Z.; Qiu, L.; Wang, Y.; Tang, S.; Fu, S. Identification and physical mapping of new PCR-Based markers specific for the long arm of rye (*Secale cereale* L.) chromosome 6. *J. Genet. Genom.* **2016**, *43*, 209–216. [CrossRef]

19. Xiao, Z.; Tang, S.; Qiu, L.; Tang, Z.; Fu, S. Oligonucleotides and ND-FISH displaying different arrangements of tandem repeats and identification of *Dasypyrum villosum* chromosomes in wheat backgrounds. *Molecules* **2017**, *22*, 973. [CrossRef]
20. Tang, Z.; Yang, Z.; Fu, S. Oligonucleotides replacing the roles of repetitive sequences pAs1, pSc119.2, pTa-535, pTa71, CCS1, and pAWRC.1 for FISH analysis. *J. Appl. Genet.* **2014**, *55*, 313–318. [CrossRef]
21. Fu, S.; Chen, L.; Wang, Y.; Li, M.; Yang, Z.; Qiu, L.; Yan, B.; Ren, Z.; Tang, Z. Oligonucleotide probes for ND-FISH analysis to identify rye and wheat chromosomes. *Sci. Rep.* **2015**, *5*, 10552. [CrossRef]
22. Han, F.; Lamb, J.C.; Birchler, J.A. High frequency of centromere inactivation resulting in stable dicentric chromosomes of maize. *Proc. Natl. Acad. Sci. USA* **2006**, *103*, 3238–3243. [CrossRef] [PubMed]
23. Untergasser, A.; Cutcutache, I.; Koressaar, T.; Ye, J.; Faircloth, B.C.; Remm, M.; Rozen, S.G. Primer3—New capabilities and interfaces. *Nucleic Acids Res.* **2012**, *40*, e115. Available online: http://bioinfo.ut.ee/primer3 (accessed on 11 May 2016). [CrossRef] [PubMed]
24. Qiu, L.; Tang, Z.X.; Li, M.; Fu, S.L. Development of new PCR-based markers specific for chromosome arms of rye (*Secale cereale* L.). *Genome* **2016**, *59*, 159–165. [CrossRef]
25. Blake, V.C.; Woodhouse, M.R.; Lazo, G.R.; Odell, S.G.; Wight, C.P.; Tinker, N.A.; Wang, Y.; Gu, Y.Q.; Birkett, C.L.; Jannink, J.L.; et al. GrainGenes: Centralized Small Grain Resources and Digital Platform for Geneticists and Breeders. *Database (Oxford)* **2019**, *2019*, baz065. Available online: https://wheat.pw.usda.gov/cgi-bin/seqserve/blast_rye.cgi (accessed on 16 August 2019). [PubMed]
26. Bauer, E.; Schmutzer, T.; Barilar, I.; Mascher, M.; Gundlach, H.; Martis, M.M.; Twardziok, S.O.; Hackauf, B.; Gordillo, A.; Wilde, P.; et al. Towards a whole-genome sequence for rye (*Secale cereale* L.). *Plant J.* **2017**, *89*, 853–869. [CrossRef]
27. Wan, A.; Wang, X.; Kang, Z.; Chen, X. Variability of the Stripe Rust Pathogen. In *Stripe Rust*; Chen, X., Kang, Z., Eds.; Springer: Dordrecht, The Netherlands, 2017; Chapter 2, pp. 35–154.
28. Hao, M.; Luo, J.; Fan, C.; Yi, Y.; Zhang, L.; Yuan, Z.; Ning, S.; Zheng, Y.; Liu, D. Introgression of powdery mildew resistance gene Pm56 on rye chromosome arm 6RS into wheat. *Front. Plant Sci.* **2018**, *9*, 1040. [CrossRef]
29. Camacho, M.V.; Matos, M.; Gonzales, C.; Perez-Flores, V.; Pernauta, B.; Pinto-Carnida, O.; Benito, C. Secale cereale inter-microsatellites (SCIMs): Chromosomal location and genetic inheritance. *Genetica* **2005**, *123*, 303–311. [CrossRef]
30. Tomita, M.; Okutani, A.; Beiles, A.; Nevo, E. Genomic, RNA, and ecological divergences of the Revolver transposon-like multi-gene family in Triticeae. *BMC Evol. Biol.* **2011**, *11*, 269. [CrossRef]
31. Xu, H.; Yin, D.; Li, L.; Wang, Q.; Li, X.; Yang, X.; An, D. Development and application of EST-based markers specific for chromosome arms of rye (*Secale cereale* L.). *Cytogenet. Genome Res.* **2012**, *136*, 220–228. [CrossRef]
32. Li, J.; Endo, T.R.; Saito, M.; Ishikawa, G.; Nakamura, T.; Nasuda, S. Homoeologous relationship of rye chromosome arms as detected with wheat PLUG markers. *Chromosoma* **2013**, *122*, 555–564. [CrossRef]
33. Friebe, B.; Jiang, J.; Raupp, W.J.; McIntosh, R.A.; Gill, B.S. Characterization of wheat alien translocations conferring resistance to diseases and pests: Current status. *Euphytica* **1996**, *91*, 59–87. [CrossRef]
34. Ren, T.H.; Yang, Z.J.; Yan, B.J.; Zhang, H.Q.; Fu, S.L.; Ren, Z.L. Development and characterization of a new 1BL. 1RS translocation line with resistance to stripe rust and powdery mildew of wheat. *Euphytica* **2009**, *169*, 207–213. [CrossRef]
35. Fu, S.; Tang, Z.; Ren, Z.; Zhang, H. Transfer to wheat (*Triticum aestivum*) of small chromosome segments from rye (*Secale cereale*) carrying disease resistance genes. *J. Appl. Genet.* **2010**, *51*, 115–121. [CrossRef] [PubMed]
36. Lei, M.P.; Li, G.R.; Liu, C.; Yang, Z.J. Characterization of wheat-Secale africanum introgression lines reveals evolutionary aspects of chromosome 1R in rye. *Genome* **2012**, *55*, 765–774. [CrossRef] [PubMed]
37. Li, Z.; Ren, Z.; Tan, F.; Tang, Z.; Fu, S.; Yan, B.; Ren, T. Molecular cytogenetic characterization of New Wheat-Rye 1R(1B) Substitution and Translocation Lines from a Chinese *Secale cereal* L. Aigan with resistance to stripe rust. *PLoS ONE* **2016**, *11*, e0163642. [CrossRef] [PubMed]

38. Lei, M.P.; Li, G.R.; Zhou, L.; Li, C.H.; Liu, C.; Yang, Z.J. Identification of wheat-*Secale africanum* chromosome 2Rafr introgression lines with novel disease resistance and agronomic characteristics. *Euphytica* **2013**, *194*, 197–205. [CrossRef]
39. Schneider, A.; Rakszegi, M.; Molnár-Láng, M.; Szakács, É. Production and cytomolecular identification of new wheat-perennial rye (*Secale cereanum*) disomic addition lines with yellow rust resistance (6R) and increased arabinoxylan and protein content (1R, 4R, 6R). *Theor. Appl. Genet.* **2016**, *129*, 1045–1059. [CrossRef]

Article

Genome-Wide Distribution of Novel Ta-3A1 Mini-Satellite Repeats and Its Use for Chromosome Identification in Wheat and Related Species

Tao Lang [1], Guangrong Li [1,*], Zhihui Yu [1], Jiwei Ma [1], Qiheng Chen [1], Ennian Yang [2] and Zujun Yang [1,*]

1 Center for Informational Biology, School of Life Science and Technology, University of Electronic Science and Technology of China, Chengdu 611731, China; langtao123xxx@126.com (T.L); 201621090111@std.uestc.edu.cn (Z.Y.); 201311090206@std.uestc.edu.cn (J.M.); 201721090218@std.uestc.edu.cn (Q.C.)

2 Crop Research Institute, Sichuan Academy of Agricultural Sciences, Chengdu 610066, China; yangennian@126.com

* Correspondence: ligr28@uestc.edu.cn (G.L.); yangzujun@uestc.edu.cn (Z.Y.); Tel.: +86-28-8320-6556 (Z.Y.)

Received: 14 January 2019; Accepted: 28 January 2019; Published: 29 January 2019

Abstract: A large proportion of the genomes of grasses is comprised of tandem repeats (TRs), which include satellite DNA. A mini-satellite DNA sequence with a length of 44 bp, named Ta-3A1, was found to be highly accumulated in wheat genome, as revealed by a comprehensive sequence analysis. The physical distribution of Ta-3A1 in chromosomes 3A, 5A, 5B, 5D, and 7A of wheat was confirmed by nondenaturing fluorescence in situ hybridization (ND-FISH) after labeling the oligonucleotide probe. The analysis of monomer variants indicated that rapid sequence amplification of Ta-3A1 occurred first on chromosomes of linkage group 5, then groups 3 and 7. Comparative ND-FISH analysis suggested that rapid changes occurred in copy number and chromosomal locations of Ta-3A1 among the different species in the tribe Triticeae, which may have been associated with chromosomal rearrangements during speciation and polyploidization. The labeling and subsequent use of Ta-3A1 by ND-FISH may assist in the precise identification and documentation of novel wheat germplasm engineered by chromosome manipulation.

Keywords: fluorescence in situ hybridization; mini-satellite; tandem repeats; wheat

1. Introduction

Repetitive DNA often constitutes the majority of grass genomes, but these repeated segments vary extensively from species to species in absolute amounts, sequences, and dispersion patterns [1]. Due to the rapid evolution and characteristic genomic distribution patterns, repetitive sequences have been useful for phylogenetic analysis and chromosomal identification, as well as for studying nuclear architecture [2,3]. According to the length of the repeated units and array size, tandem repeated DNA sequences can be classified into three groups, including microsatellites, mini-satellites, and satellite DNA. The mini-satellites (10–60 bp) were first described by Jeffreys et al. [4], and are usually defined as the repetition in tandem of a short (6- to 100-bp) motif spanning 0.5 kb to several kilobases [5,6]. Mini-satellites were first characterized in the human genome [4], where they share a common motif known as the core sequence [7], and were subsequently discovered in a variety of plant species [8–11]. A remarkable advance in the knowledge of repetitive sequences has occurred in recent years because of the introduction of next-generation sequencing technologies [12]. Bioinformatic pipelines or web servers have provided computational facilities for the identification of mini-satellites for next-generation sequence data [13–15]. The availability of genome sequences allows the development of site-specific mini-satellites, as has been reported for diploid and polyploid plant species [16–19].

Bread wheat, *Triticum aestivum* L., consists of three closely related subgenomes (A, B, D), and these genomes contain a high proportion (>85%) of repetitive DNA [20,21]. Universal repetitive sequences and genome-specific repetitive sequence families have been identified in Triticeae species [22]. The mini-satellite sequences in wheat and related species have been referred to as "variable number tandem repeats" (VNTRs). The identified mini-satellite sequences in *T. aestivum*, *T. durum* Desf., *T. monococcum* L., *Aegilops speltoides* Tausch, and *Ae. tauschii* Coss. have shown a high degree of polymorphism among the various genomes [23,24]. Probes and markers for mini-satellites have been successfully used for DNA fingerprinting [11]. Extensive analysis of such mini-satellite tandem repeats is potentially useful for combining genomic and cytogenetic studies of wheat chromosomes.

The recent completion of the genomic sequencing of *T. aestivum* cv. Chinese Spring (International Wheat Genome Sequencing Consortium, [25]), and also the A genome donor (*Triticum urartu* Thumanjan ex Gandilyan) [26], the D genome progenitor (*Ae. tauschii*) [27], and the tetraploid wheat, *Triticum dicoccoides* Schrank ex Schübl. (AABB genomes [28]), provide opportunities for the localization of repetitive sequences to chromosome physical regions. Lang et al. [29] first constructed a database of repeats according to unit size, array number, and physical coverage length of TRs in the wheat genome. The comparison of tandem repeats, including satellite repeats, at genome-wide levels may potentially shed new light on our understanding of genomic amplification through evolutionary processes [3,29,30].

The principal aim of the present study is to describe the physical location and sequence variations of a novel mini-satellite repeat, Ta-3A1, between wheat and its ancestors by genome-wide analysis. Fluorescence in situ hybridization by Ta-3A1 for chromosome identification and evolution of wheat and related species is also discussed.

2. Materials and Methods

2.1. Plant Materials

Common wheat cv. Chinese Spring (CS), *Thinopyrum intermedium* (Host) Barkworth & D.R. Dewey, and *Th. ponticum* (Podp.) Z.-W.Liu & R.-C.Wang were provided by Prof. Bikram S. Gill, Wheat Genetic and Genomic Resource Center, Kansas State University. Germplasm of *Aegilops* L. and *Hordeum* L. species were obtained from Prof. Eviatar Nevo at the Institute of Evolution, University of Haifa, Israel. The durum wheat—*Dasypyrum villosum* (L.) Borbás amphiploid was obtained from Prof. Huaren Jiang, Sichuan Agricultural University, China. Lines of tetraploid wheat, including *T. carthlicum* Nevski, *T. abyssinicum* L., *T. durum*, and *T. dicoccoides*, were provided by Dr. Cheng Liu, Crop Research Institute, Shandong Academy of Agricultural Sciences. Tetraploid *Ae. tauschii* (Zeng et al. 2012) [31], rye (*Secale cereale* L.), *D. villosum* L., hexaploid triticale cv. Currency, and the wheat—*D. breviaristatum* L. partial amphiploid [32] were maintained in our laboratory at the School of Life Science and Technology, University of Electronic Science and Technology of China. A list of accessions with their ploidy level and genome constitution are given in Table S1.

2.2. Identification of Tandem Repeats (TRs)

The wheat genome IWGSC RefSeq v1.0 assembly and genomic annotation was downloaded from (https://wheat-urgi.versailles.inra.fr/). The genome-wide analysis of the tandem repeats in wheat was conducted by the Tandem Repeat Finder (TRF) algorithm [33] using alignment parameters of 2, 7, and 7 for match, mismatch, and indels, respectively. The interpretation of TRs by number of copy, array, and cluster is illustrated in Figure S1. Tandem repeats annotated by TRF were divided into three classes according to the size of period distances (<20, 20–60, and >60). A nonredundant set of sequences was constructed using the program CD-HIT [34] at an 85% similarity level. The distribution and chromosomally enriched locations of the clusters of TRs in wheat genome were visualized with the assistance of a website, B2DSC (http://mcgb.uestc.edu.cn/b2dsc), according to the procedure of Lang et al. [29].

2.3. Fluorescence In Situ Hybridization (FISH)

Seedling roots from wheat and related species and wheat-alien derivatives were collected when they were about 2–3 cm long. Excised root tips were treated with nitrous oxide gas for 2 h under 1.0 MPa pressure. The treated root tips were fixed in 90% acetic acid for 10 min. After the roots were washed by distilled water, they were digested in 2% cellulase and 1% pectolyase (Yakult Pharmaceutical, Tokyo, Japan) for 55 min at 37 °C. The digested root sections were washed and then mashed to form a cellular suspension in 100% acetic acid. The cell suspensions were dropped onto glass slides for chromosome preparation according to Kato et al. [35]. The synthesized oligonucleotide probes Oligo-pSc119.2 with 5′ labeled by 6-carboxyfluorescein (6-FAM) and Oligo-pTa535 with 5′ labeled by 6-carboxytetramethylrhodamine (Tamra) were used for identifying the wheat chromosomes according to the description of Tang et al. [36]. The oligonucleotide probes were synthesized by Shanghai Invitrogen Biotechnology Co. Ltd. (Shanghai, China). The protocol of nondenaturing FISH (ND-FISH) employing synthesized probes was described by Fu et al. [37]. After ND-FISH, the chromosome squashes for sequential FISH for Oligo-3A1 labeled by 6-FAM were washed twice, each for 5 min with 0.1% Tween 20 in 2 × SSC, to remove the hybridization signals. FISH images were captured with an Olympus BX-51 microscope equipped with a DP-70 CCD camera operated by DP manager software (Olympus). Images were processed using Photoshop 3.0 (Adobe Systems Incorporated, San Jose, CA, USA).

2.4. Sequences Variation and Phylogenetic Analysis

The DnaSP version 6.0 [38] was used to compute DNA polymorphism and their variances. Multiple sequence alignments from different chromosomal regional accumulated Oligo-3A1 sequences were done using ClustalX [39] and were subsequently used for phylogenetic analyses. Maximum likelihood (ML) trees were constructed by MEGA7.0 [40], and the trees were displayed with FigTree v1.40 software (http://tree.bio.ed.ac.uk/software/figtree/). Clustered consensus sequences combined with the phylogenetic trees were displayed with the logos [41,42].

3. Results

3.1. Ta-3A1 Was One of the Top Accumulated TRs

A database of all the raw and filtered nonredundant TRs predicted for the IWGSC RefSeq v1.0 assembly by TRF is available at http://mcgb.uestc.edu.cn/tr [29]. The nonredundant TRs were classified into three classes based on their lengths of period distance (PD), specifically <20 bp, 20–60 bp, and >60 bp. The distribution of different PD sizes on individual chromosomes indicated that the copy numbers of TRs of <20 bp were more abundant than those with 20–60 bp and >60 bp (Figure 1). Interestingly, the TRs of 20–60 bp were highly accumulated within short regions of chromosomes 3A, 5A, 7A, 5B, and 5D. In order to determine the composition of the highly accumulated mini-satellite repeats in the five chromosomes, the TRs with a pattern size ≥20 bp, copy number ≥50, and percent match ≥80 were grouped. We found that a TR array with a length of 44 bp predominated (the consensus sequence was AATAATTTTACACTAGAGTTGAACTAGCTCTA TAAGCTAGTTCA). This predominating array was named TR-3A1 and consisted of 280 nonredundant arrays covering a length of 2,630,397 bp. Across all wheat chromosomes, the estimated total predicted copy number of TR-3A1 was 59,801, which was mainly distributed on the above five chromosomes with copy numbers of 14,987, 11,083, 7828, 8444, and 28,964 on chromosomes 3A, 5A, 7A, 5B, and 5D, respectively (Table 1).

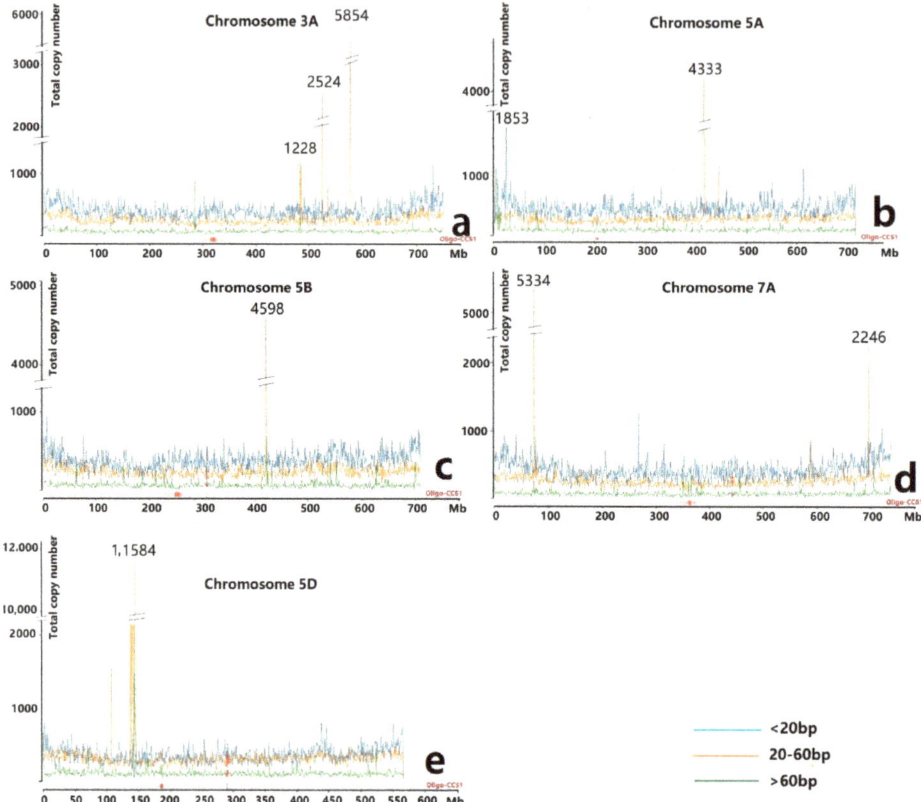

Figure 1. The copy number and distribution of tandem repeats in wheat chromosomes 3A (**a**), 5A (**b**), 5B (**c**), 7A (**d**), and 5D (**e**). The copy number was predicted for wheat Chinese Spring genomic sequences v 1.0 at a web server at http://mcgb.uestc.edu.cn/BD2SC with default parameters.

Table 1. The predicted copy numbers of Ta-3A1 on sequenced *Triticum urartu*, *Aegilops tauschii*, *T. dicoccoides*, and *T. aestivum* genomes.

Species	Chromosome	Copy Number	Species	Chromosome	Copy Number
	3A	14,987	*T. urartu*	3A	2304
	5A	11,083		5A	1199
T. aestivum	7A	7828		7A	835
	5B	8444	*T. dicoccoides*	3A	2765
	5D	28,964		5A	16,482
				7A	6449
Ae. tauschii	5D	5404		5B	2200

The Ta-3A1 sequence was submitted as a query into the website B2DSC using default parameters (pident = 85 and qcovhsp = 80). An example of the entire genomic distribution, detailed chromosomal locations, as well as the sequence directional analysis of Ta-3A1 according to the website B2DSC appears in Figure 2. The predicted copy number and the estimated physical positions of Ta-3A1 (Figure 2a) resembled the 20–60 bp TR localizations in Figure 1, which suggested that Ta-3A1 was actually the principal predicted mini-satellite in wheat genome compared to the results of TR enrichment studies by Lang et al. [29]. The physical organization of Ta-3A1 on chromosome 5DS regions at 146–147 Mb

(Figure 2b) and the head-to-head repeats of the 72.19 Mb region in chromosome 7A (Figure 2c) were indicated in detail.

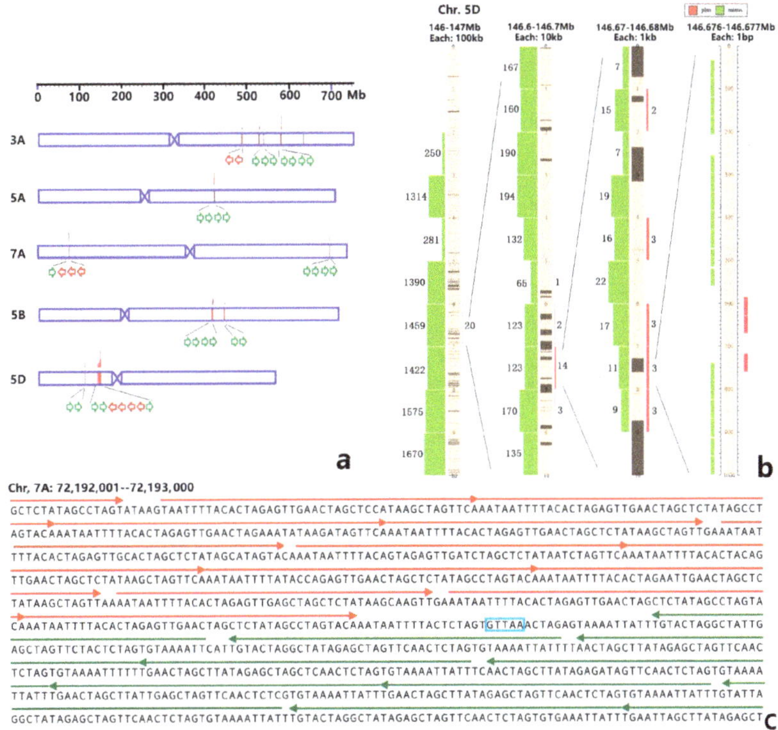

Figure 2. The predicted physical location of Ta-3A1 in wheat genome. The overview of Ta-3A1 on five chromosomes (**a**), copy number in a region of 5D (**b**), and an array of sequences on 7A (**c**) were indicated, respectively. The arrows represent the mini-satellite array, and green and red arrows show the minute and plus strand, respectively.

Similarly, by using the B2DSC website, the copy numbers and physical locations of Ta-3A1 were also predicted for the sequenced genomes of wheat's ancestral species, including the A genome of *T. urartu* [26], the D genome [27], and AABB genomes [28] (Table 1). Based on the comparison of overall genomic copy numbers of Ta-3A1 among those of the A, B, and D genomes, we found that the copy numbers on each of the corresponding chromosomes differed significantly, and the copy numbers increased from diploid to tetraploid to hexaploid wheat. The most abundant sites of Ta-3A1 were located on 5D of wheat, which was significantly higher than those on 5D of *Ae. tauschii*. It is likely that the copy numbers of Ta-3A1 largely increased during the polyploidization of wheat.

3.2. Sequence Variability of Ta-3A1 on Different Chromosomal Regions

The monomer sequences within each of the Ta-3A1 arrays spanning over 200 copies were aligned according to their diagnostic physical positions. A total of 17 clustered physical regions among the 3A, 5A, 7A, 5B, and 5D chromosomes, which represented from 17% to 43% of the nonredundant sequences, were analyzed for sequence variation rates (Table 2). Interestingly, despite their different degrees of abundance, mean π values of Ta-3A1 monomers for each of the chromosome regions were roughly similar (from 9.3% in 5B to 19.4% in 3A). The higher nucleotide diversity values (π) on different chromosomes and different regions may suggest that the sequence variation and rearrangements of

Ta-3A1 were due to recombination events, which may be an important force generating new monomers in chromosomes 3A, 5D, and 7A.

Table 2. Nucleotide polymorphism of the repeat features of Ta-3A1 in wheat genome.

Chromosome	Physical Location (bp)	*N*	*S*	π	Hd ± SD	Tajima's *D*
3A	484,160,160–484,477,359	674	16	0.131	0.925 ± 0.007	−1.865 *
	486,643,172–486,682,396	370	24	0.147	0.987 ± 0.002	−1.851 *
	525,926,313–525,999,776	356	22	0.127	0.965 ± 0.005	−2.014 *
	526,002,472–526,436,842	1065	17	0.150	0.959 ± 0.004	−1.648
	536,013,749–536,065,681	280	22	0.144	0.981 ± 0.003	−1.991 *
	577,811,191–577,988,203	685	24	0.155	0.991 ± 0.001	−1.719
	578,013,600–578,743,842	2821	5	0.194	0.713 ± 0.006	−0.916
5A	420,664,083–420,976,973	4031	4	0.159	0.520 ± 0.010	−0.998
5B	412,801,830–414,820,389	2806	9	0.130	0.765 ± 0.007	−1.456
	441,918,962–442,205,867	616	22	0.127	0.973 ± 0.004	−1.920 *
5D	109,229,486–109,417,074	908	19	0.129	0.948 ± 0.005	−1.843 *
	139,568,074–144,410,834	3529	3	0.093	0.263 ± 0.010	−1.276
	146,286,068–146,411,073	762	21	0.149	0.980 ± 0.003	−1.765
	146,494,968–147,180,961	2924	9	0.119	0.693 ± 0.010	−1.519
	147,436,033–147,525,419	570	22	0.143	0.983 ± 0.003	−1.883 *
7A	72,124,011–72,576,573	1466	14	0.120	0.866 ± 0.008	−1.743
	695,384,113–695,559,280	621	22	0.128	0.970 ± 0.004	−1.942 *

Number of monomeric repeats sequenced (*N*), number of variable sites (*S*), nucleotide diversities (π), and haplotype diversity (Hd). Standard deviation (SD), The estimates of the number of segregating sites and the average number of nucleotide differences correlated under the neutral model (Tajima's D). * showed significant difference at $p < 0.05$.

In order to reveal the evolutionary aspects of the overall Ta-3A1 monomer sequence variation, a phylogenetic tree of the representative Ta-3A1 satDNA in all the chromosome regions was constructed. A total of 216 representative Ta-3A1 monomer sequences predicted from the wheat genome were used for constructing the phylogenetic tree (Figure S2 and Figure 3). The results suggested that the monomers from chromosome 5D were separated first, the sequences of 5A, 5B, and 3A were categorized as the second subgroup, and the 7A sequences appeared in the third group. The sequences of chromosome 3A were mainly clustered in groups 6, 8, and 9. Combining the phylogenetic tree with the physical locations of the Ta-3A1 sequences (Figure 3), we found that the Ta-3A1 sequences from 5D distributed in most groups: However, the sequences in 5B were mainly concentrated on groups 3 and 4. This suggests that the Ta-3A1 on 5D evolved several times during sequence expansion and became more active than that of the 5B chromosome.

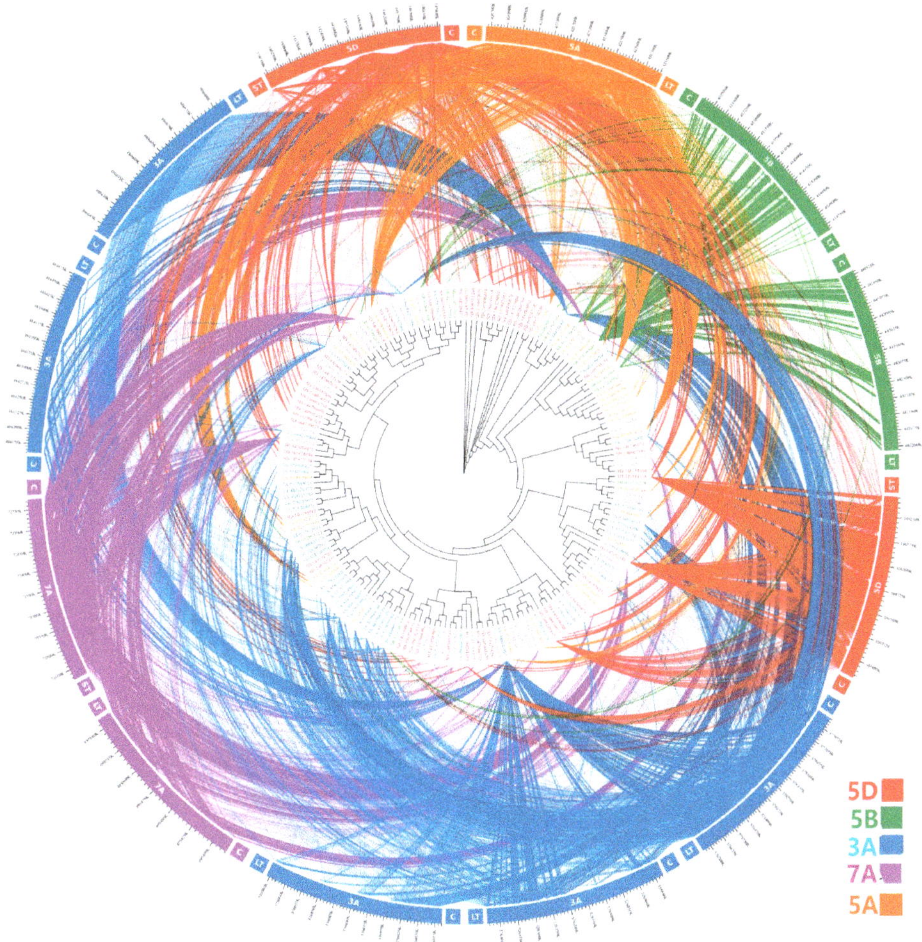

Figure 3. Phylogenetic tree of Ta-3A1 repeats and their physical location on five wheat chromosomes.

3.3. *FISH of Oligo-3A1 in Wheat*

Based on the sequences of the Ta-3A1, we produced a labeled oligo probe (Oligo-3A1) for revealing the chromosome organization of Ta-3A1 by ND-FISH analysis. Mitotic metaphase chromosomes from bread wheat, durum wheat, and some related species were prepared to study the hybridization patterns of Oligo-3A1. Sequential ND-FISH with Oligo-pTa535 and Oligo-pSc119.2 was performed for the identification of individual chromosomes of wheat [36]. As shown in Figure 4, the ND-FISH hybridization patterns of Oligo-3A1 in Chinese Spring wheat displayed strong signals on pericentromeric regions on 5DS, intercalaric regions of 5AL, 3AL, and 5BL, and subtelomeric regions of both arms on 7A. The chromosomal distribution of the FISH signals were consistent with the predicted copy numbers of each chromosome (Table 1).

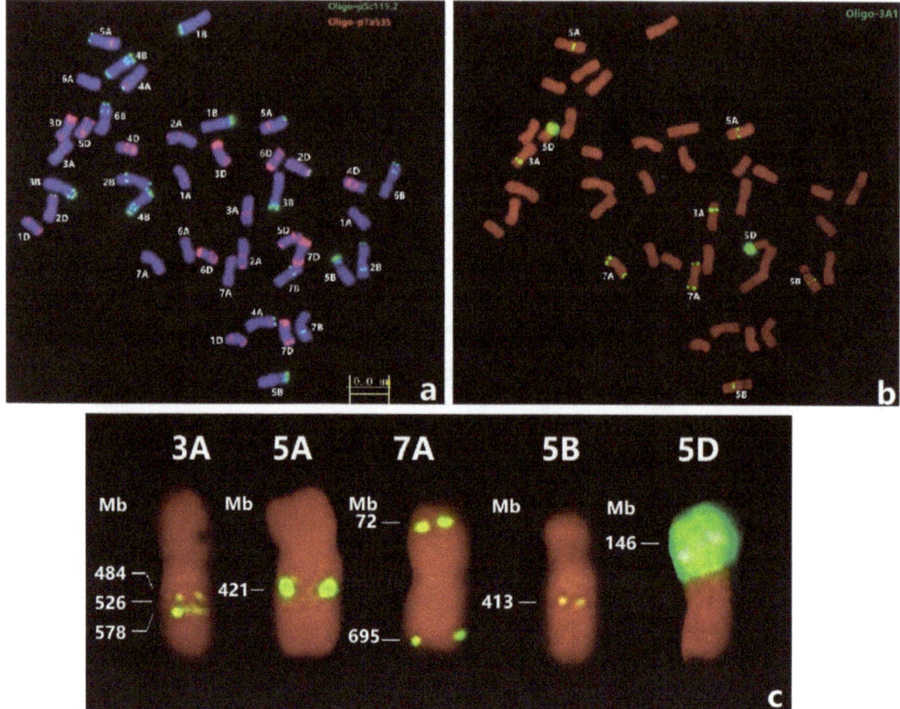

Figure 4. Nondenaturing fluorescence in situ hybridization (ND-FISH) patterns of Ta-3A1 repeats in wheat Chinese Spring: The probes Oligo-pSc119.2 (green) and pTa535 (red) (**a**), and sequential Oligo-3A1 (green) (**b**), were used for ND-FISH. The copy numbers predicted by web server page B2DSC and the copy number of Ta-3A1 at the specific region of each 1-Mb scale were illustrated in karyotypes of chromosomes (**c**). Bar indicated 10 μm.

3.4. FISH of Oligo-3A1 in Representative Triticeae Species

In order to shed light on the genetic divergence of Ta-3A1 on representing wheat ancestral species, the chromosome preparations from *T. monococcum* (A genome), *Ae. longissima*, *Ae. searsii*, *Ae. speltoides* (S genomes), and *Ae. tauschii* (D genome) accessions were carried out by ND-FISH analysis (Figure 5). The results showed that the signal strengths varied from faint to strong on $5A^m$ and $3A^m$ to $7A^m$ in *T. monococcum* (Figure 5a,d). Strong hybridization signals appeared on long arms of 5S in *Ae. speltoides* (Figure 5b,e), but conversely on the short arms of $5S^l$ in *Ae. longissima* (Figure 5c,f) and $5S^s$ in *Ae. searsii* (Figure 5h,k). Relatively strong signals of Oligo-3A1 were observed on 5DS of *Ae. tauschii* (Figure 5j,m). Therefore, the strength of signals and the localization of chromosomes varied among the wheat A, B, and D ancestral genomes.

A total of each of 12 *T. durum*, six *T. dicoccoides*, three *T. carthlicum*, and two *T. abyssinicum* accessions were studied by ND-FISH using Oligo-3A1 (Figure 5m–r, Table 3). We observed that the signals on chromosomes 3A and 7A were stable, but the strength of signals on chromosomes 5A and 5B varied extensively. Nine accessions showed strong signals on 5BL and weak signals on 5AL, and the converse was noted in two accessions. The remaining accessions had equally strong hybridization signals on 5AL and 5BL. It is interesting to note that additional discrete signals were observed on chromosome 2BL of five *T. durum* lines, which were derived from International Maize and Wheat Improvement Center (CIMMYT) (Figure 5m,p). The highly polymorphic FISH patterns of Oligo-3A1 were observed among tetraploid wheat accessions.

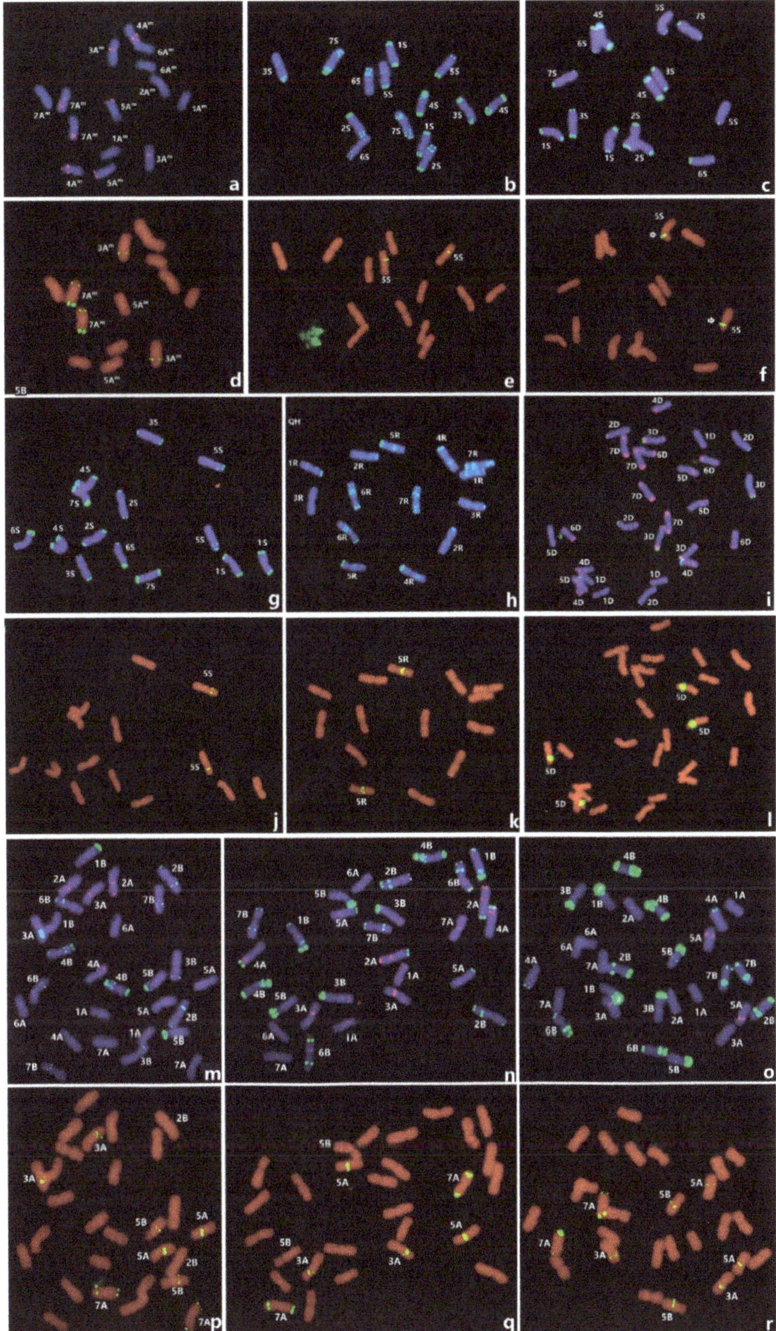

Figure 5. Hybridization of the ND-FISH probes on chromosomes of *T. monococcum* (**a**,**d**), *Ae. speltoides* (**b**,**e**), *Ae. longissima* (**c**,**f**), *Ae. searsii* (**g**,**j**), *Secale cereale* (**h**, **k**), tetraploid *Ae. tauschii* (**i**,**l**), *T. durum* (**m**,**p**), *T. carthlicum* (**n**,**q**), and *T. dicoccoides* (**o**,**r**). The probes Oligo-pSc119.2 (green) and pTa535 (red) (**a**–**c**,**g**–**i**,**o**–**p**), and sequential Oligo-3A1 (green) (**d**–**f**,**j**–**l**,**p**–**r**) were used, respectively.

Table 3. The differential hybridization patterns of Ta-3A1 among tetraploid wheat.

Species	Origin	Accessions	5B	5A	2BL
T. dicoccoides	Japan	H137	++	++	
T. dicoccoides	Israel	H141	+++	++	
T. dicoccoides	Syria	H177	++	+++	
T. dicoccoides	Syria	H179	+	++	
T. dicoccoides	China	H171	+	++	
T. dicoccoides	Israel	H283	+	+++	
T. carthlicum	Turkey	H149	+	+	
T. carthlicum	Turkey	H94	+	+	
T. carthlicum	USA	AS311	+	+++	
T. abyssinicum	Ethiopia	H129	+	+	
T. abyssinicum	Ethiopia	H131	+	+	
T. durum	CIMMYT	H281 H322 H268 H242	+	+++	+
T. durum	CIMMYT	H246	++	+++	+
T. durum	CIMMYT	H261	++	+++	++
T. durum	CIMMYT	H242	+	++	+
T. durum	CIMMYT	H259	+++	+++	++
T. durum	Israel	H319 H305 H299	+	+++	
T. durum	Israel	Langdon	+	++	

The strength of Ta-3A1 hybridization signals indicated by + (weak), ++ (medium), +++ (strong).

Additionally, ND-FISH with Oligo-3A1 was also conducted on chromosome preparations of other related wild diploid or polyploid species and wheat-alien species amphiploids. The identification of individual chromosomes from *Secale* [37], *Dasypyrum* [43], and *Thinopyrum* [44] was achieved by Oligo-pSc119.2 and Oligo-pTa535. Barley chromosomes were completely devoid of any hybridization signals of Oligo-3A1, as were *D. breviaristatum* chromosomes in the wheat—*D. breviaristatum* amphiploid. The hybridization signals of Oligo-3A1 on rye chromosome 5R (Figure 5) and in triticale (Figure 6a,b) were relatively strong, while the signals on *D. villosum* 5V were weak in the durum wheat—*D. villosum* amphiploid (Figure 6e,f). Moreover, only two chromosomes of *Th. ponticum* ($2n = 10x = 70$) appeared to carry Oligo-3A1 signals on their long arms (Figure 6c,d), and only two pairs of chromosomes with Oligo-3A1 sites (one in short arms and the other in long arms) were observed for *Th. intermedium* ($2n = 6x = 42$) (Figure 6g,h). The distribution of Ta-3A1 was polymorphic for presence/absence in different genomes or species, which suggests that the major loci were probably lost during the evolution of this polyploidy species.

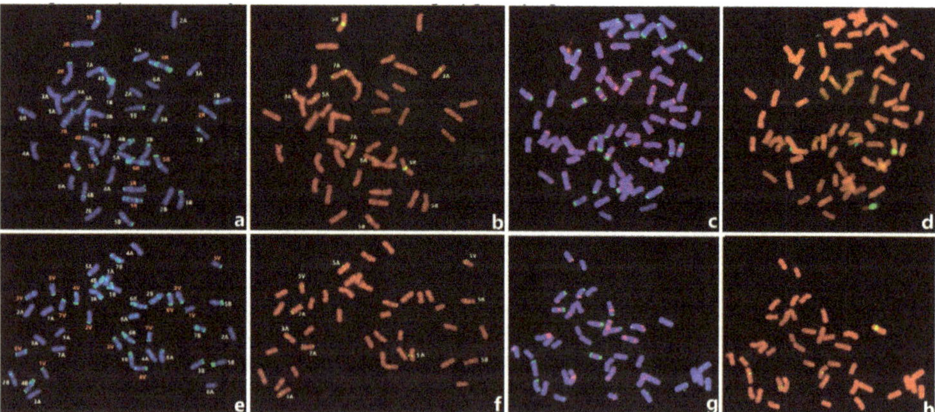

Figure 6. ND-FISH on chromosomes of hexaploid triticale (**a**,**b**), *Thinopyrum ponticum* (**c**,**d**), *T. durum–Dasypyrum villosum* amphiploid (**e**,**f**), and *Th. intermedium* (**g**,**h**). The probes Oligo-pSc119.2 and pTa535 (**a**,**c**,**e**,**g**) and sequential Oligo-3A1 (**b**,**d**,**f**,**h**) were used, respectively.

4. Discussion

With the recent development of next-generation sequencing technologies, an ever-increasing number of plant species have been studied and vast amounts of sequence data accumulated. Next-generation sequencing provides a convenient way to search for the presence of mini-satellite repeated sequences [14,15]. As a major class of repetitive DNA, mini-satellites consist of tandemly organized monomers. Zakrzewski et al. [16] isolated 517 novel repetitive sequences and used them for the identification of mini-satellite families in *Beta vulgaris*, and they revealed that mini-satellites are moderately to highly amplified by bioinformatic analysis and southern hybridization. Mogil et al. [15] identified five distinct, interleaved mini-satellite families in the pericentromeric regions of soybean (*Glycine max*) chromosomes, which were mediated by >3200 intact copies. The availability of a whole-genome assembly of wheat cv. Chinese Spring also provides opportunities to examine the distribution of retrotransposons [45] and of TRs [29] at a whole-genomic level. Our recent study predicted that TRs occupy about 3–5% of wheat chromosomes [29]. Overall, the total array number of TR mini-satellites (20–60 bp) was 1,453,457, and the total length of such mini-satellites reached 129,873,148 bp in the wheat genome by TRF [29]. Martienssen and Baulcombe [46] first identified mini-satellite arrays lying upstream of an alpha-amylase gene in hexaploid wheat. Somers et al. [24] amplified mini-satellite core sequences from wheat. However, these mini-satellite arrays in wheat represented moderately repeated families of less than 100 copies per array. Tang et al. [47] reported five probes with mini-satellite sequences among ND-FISH probes as predicted by TRF, and we predicted 120 to 260 copy numbers across specific chromosome regions with total lengths of under 20 kb by B2DSC web server. The present study found that Ta-3A1 represented the highest copy number of a mini-satellite yet reported, and it covered a total length of over 2 Mb and was mainly located on five chromosomes of wheat (Figure 1). Two copies of Ta-3A1-like sequences were found in the intron 2 of Bradi1g11210.1 and the intron 4 of Bradi3g03878.1 in the *Brachypodium* genome. Ta-3A1 was absent from the sequenced barley genome [48]. With regard to the accumulation of Ta-3A1 in pericentromeric regions of chromosome 5D, the observed high number may have been related to centromeric specific retrotransposons (Figure 7). Just how the extensive duplication of the Ta-3A1 sequence on 5D occurred, which resulted in such great copy numbers, is as yet unexplained and needs to be further investigated. A problem exists with the identification of the short tandem repeats, which are often reshuffled and diversified and are somehow able to escape computational detection. The quality of assembly of the wheat genome needs to be improved through the use of single-molecule sequencing, optical mapping, and chromosome conformation capture technologies [49,50]. The updated assembly of wheat genome will offer unprecedented insights into the detection of the structural features and relevant evolutionary characteristics of TRs, including mini-satellites.

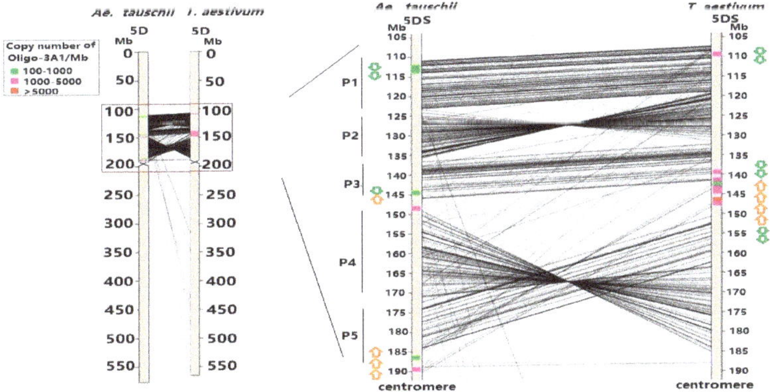

Figure 7. The genome collinearity of pericentromeric regions of chromosome 5D between *Ae. tauschii* and wheat Chinese Spring (CS).

The sequence dynamics of a specific satDNA family may differ across genomic regions [51], populations [52], species, or higher taxonomic groups [53–56]. Consequently, detailed characterization of satDNA families has revealed that evolutionary patterns are more complex than previously anticipated. The study of mini-satellite sequence elements is essential to our understanding of the nature and consequences of genome size variation between different species and for studying the large-scale organization and evolution of grass genomes [5]. With regard to the ancestral species of *Hordeum* separated by a time period ranging from 11.6 to 15.6 Mya [48], the Ta-3A1 sequence might not have been amplified at all since no signals were detectable in wild and cultivated barley species. The *D. breviaristatum* species was also devoid of Ta-3A1 hybridization sites. It is likely that *D. breviaristatum* may be one of the ancestral Triticeae species after barley separation, which is consistent with data from our previous studies on seed gliadin evolution [57]. Bread wheat can strongly adapt to different climates, and one of the key factors of this characteristic is the allohexaploid genome structure, which originates from two distinct polyploidization events [58]. In the present study, we found that the copy number of Ta-3A1 was largely amplified from *T. urartu* (AA), *Ae. tauschii* (DD), and allotetraploid *T. turgidum* (AABB) to common wheat (AABBDD). Moreover, *Ae. speltoides* is cytogenetically distinct from the S genomes of other diploid and polyploid *Aegilops* species based on C-banding and FISH [59,60]. We also showed that the FISH patterns of Ta-3A1 appeared on the long arm of 5S in *Ae. speltoides* and showed close resemblance to those of the B genome in wheat. Therefore, the study of the accumulation of mini-satellite sequences can shed light on wheat genome organization and, specifically, on the role of repetitive elements by using Ta-3A1 as an example.

The satDNA family is characterized by complex features, including large variations in copy number and long-range organization of repeat units, genome location and distribution, as well as interchromosomal and intrachromosomal recombination rates [53,61–65]. Fluorescence in situ hybridization (FISH) using the probes of repetitive DNA sequences was used to determine their physical locations on individual chromosomes of common wheat and its relatives [66,67]. In this current study, we also found that *Secale*, *Dasypyrum*, and *Thinopyrum* species displayed varied FISH patterns of Ta-3A1 hybridization (Figure 6). It is possible to deduce trends in the complexity of Ta-3A1 organization during the evolution of the Triticeae tribe from wild species to cultivated species. The variability of Ta-3A1 representing mini-satellites has been used in numerous studies to identify plant chromosomes. The Ta-3A1 probe has been used to precisely identify the chromosomal breakage points in wheat—*Th. intermedium* introgression lines [68,69]. FISH specific hybridization sites by Ta-3A1 on *Secale*, *Dasypyrum*, and *Aegilops* chromosomes can also trace specific chromatin in wheat-alien introgression lines. The combination of both molecular cytogenetics and genomic research on TRs, including mini-satellites, will significantly benefit future wheat breeding activities, which focus on chromosome manipulation and engineering [29].

5. Conclusions

The mini-satellite sequences are abundantly distributed in the wheat genome, which served as powerful cytogenetic and genomic markers for germplasm enhancement and breeding practices. The present study identified a novel mini-satellite repeat Ta-3A1, and revealed its copy number and physical locations of Ta-3A1 among chromosomes of wheat and their ancestral genomes. The large sequence variation of Ta-3A1 displayed rapid evolution of mini-satellites occurred during speciation and polyploidization The Ta-3A1 based ND-FISH may be helpful for chromosome identification of wheat and related species, which can be directly used to wheat improvement by chromosome engineering.

Supplementary Materials: The following are available online at http://www.mdpi.com/2073-4395/9/2/60/s1, Table S1: List of materials and their origins; Figure S1: The interpretation of TRs by numbers of individual copy, array, and cluster; Figure S2: Phylogenetic tree and sequence comparison of the Ta-3A1 repeats.

Author Contributions: Z.Y. (Zujun Yang) and G.L. designed the experiments. T.L., G.L., Z.Y. (Zhihui Yu), J.M. and Q.C. performed the experiments. Z.Y. (Zujun Yang), E.Y., and T.L. analyzed the data. Z.Y. (Zujun Yang) and G.L. wrote the paper.

Funding: This research was funded by the National Key Research and Development Program of China (2016YFD0102000), the Applied and Basic Project (2016JY0075) from the Science and Technology Department of Sichuan, China, and the National Natural Science Foundation of China (No. 31171542).

Acknowledgments: We are thankful to Ian Dundas at the University of Adelaide, Australia, for the helpful comments on the manuscript.

Conflicts of Interest: The authors declare no conflict of interest.

References

1. Schmidt, T.; Heslop-Harrison, J.S. Genomes, genes and junk: The largescale organization of plant chromosomes. *Trends Plant Sci.* **1998**, *3*, 195–199. [CrossRef]
2. Heslop-Harrison, J.S. Comparative genome organization in plants: From sequence and markers to chromatin and chromosomes. *Plant Cell* **2000**, *12*, 617–636. [CrossRef] [PubMed]
3. Contento, A.; Heslop-Harrison, J.S.; Schwarzacher, T. Diversity of a major repetitive DNA sequence in diploid and polyploid Triticeae. *Cytogenet. Genome Res.* **2005**, *109*, 34–42. [CrossRef] [PubMed]
4. Jeffreys, A.J.; Wilson, V.; Thein, S.L. Hypervariable "minisatellite" regions in human DNA. *Nature* **1985**, *314*, 67–73. [CrossRef] [PubMed]
5. Vergnaud, G.; Denoeud, F. Minisatellites: Mutability and genome architecture. *Genome Res.* **2000**, *10*, 899–907. [CrossRef] [PubMed]
6. Mehrotra, S.; Goyal, V. Repetitive sequences in plant nuclear DNA: Types, distribution, evolution and function. *Genom. Proteom. Bioinform.* **2014**, *12*, 164–171. [CrossRef]
7. Wong, Z.; Wilson, V.; Patel, I.; Povey, S.; Jeffreys, A.J. Characterization of a panel of highly variable minisatellites cloned from human DNA. *Ann. Hum. Genet.* **1987**, *51*, 269–288. [CrossRef]
8. Rogstad, S.H.; Patton, J.C.; Schaal, B.A. M13 repeat probe detects DNA minisatellite-like sequences in gymnosperms and angiosperms. *Proc. Natl. Acad. Sci. USA* **1988**, *85*, 9176–9178. [CrossRef]
9. Broun, P.; Tanksley, S.D. Characterization of tomato DNA clones with sequence similarity to human minisatellites 33.6 and 33.15. *Plant Mol. Biol.* **1993**, *23*, 231–242. [CrossRef]
10. Tourmente, S.; Deragon, J.M.; Lafleuriel, L.; Tutois, S.; Pélissier, T.; Cuvillier, C.; Espagnol, M.C.; Picard, G. Characterization of minisatellites in Arabidopsis thaliana with sequence similarity to the human minisatellite core sequence. *Nucleic Acids Res.* **1994**, *22*, 3317–3321. [CrossRef]
11. Zhou, Z.; Bebeli, P.J.; Somers, D.J.; Gustafson, J.P. Direct amplification of minisatellite-region DNA with VNTR core sequences in the genus *Oryza*. *Theory Appl. Genet.* **1997**, *95*, 942–949. [CrossRef]
12. Treangen, T.J.; Salzberg, S.L. Repetitive DNA and next-generation sequencing: Computational challenges and solutions. *Nat. Rev. Genet.* **2011**, *13*, 36–46. [CrossRef] [PubMed]
13. Bally, P.; Grandaubert, J.; Rouxel, T.; Balesdent, M.H. FONZIE: An optimized pipeline for minisatellite marker discovery and primer design from large sequence data sets. *BMC Res. Notes* **2010**, *3*, 322. [CrossRef] [PubMed]
14. Abouelhoda, M.; El-Kalioby, M.; Giegerich, R. WAMI: A web server for the analysis of minisatellite maps. *BMC Evol. Biol.* **2010**, *10*, 167. [CrossRef] [PubMed]
15. Mogil, L.S.; Slowikowski, K.; Laten, H.M. Computational and experimental analyses of retrotransposon-associated minisatellite DNAs in the soybean genome. *BMC Bioinform.* **2012**, *13*, S13. [CrossRef] [PubMed]
16. Zakrzewski, F.; Wenke, T.; Holtgräwe, D.; Weisshaar, B.; Schmidt, T. Analysis of a c0t-1 library enables the targeted identification of minisatellite and satellite families in Beta vulgaris. *BMC Plant Biol.* **2010**, *10*, 8. [CrossRef] [PubMed]
17. Honma, Y.; Yoshida, Y.; Terachi, T.; Toriyama, K.; Mikami, T.; Kubo, T. Polymorphic minisatellites in the mitochondrial DNAs of Oryza and Brassica. *Curr. Genet.* **2011**, *57*, 261–270. [CrossRef] [PubMed]
18. Oliveira, G.A.F.; Dantas, J.L.L.; Oliveira, E.J. Development and validation of minisatellite markers for Carica papaya. *Biol. Plant.* **2015**, *59*, 686–694. [CrossRef]
19. Orzechowska, M.; Majka, M.; Weiss-Schneeweiss, H.; Kovařík, A.; Borowska-Zuchowska, N.; Kolano, B. Organization and evolution of two repetitive sequences, 18-24J and 12-13P, in the genome of *Chenopodium* (*Amaranthaceae*). *Genome* **2018**, *61*, 643–652. [CrossRef] [PubMed]

20. Brenchley, R.; Spannagl, M.; Pfeifer, M.; Barker, G.L.A.; D'Amore, R.; Allen, A.M.; McKenzie, N.; Kramer, M.; Kerhornou, A.; Bolser, D.; et al. Analysis of the bread wheat genome using whole-genome shotgun sequencing. *Nature* **2012**, *491*, 705–710. [CrossRef] [PubMed]
21. Zimin, A.V.; Puiu, D.; Hall, R.; Kingan, S.; Clavijo, B.J.; Salzberg, S.L. The first near-complete assembly of the hexaploid bread wheat genome, *Triticum aestivum*. *Gigascience* **2017**, *6*, 1–7. [CrossRef]
22. Anamthawat-Jónsson, K.; Heslop-Harrison, J.S. Isolation and characterization of genome-specific DNA sequences in Triticeae species. *Mol. Gen. Genet.* **1993**, *240*, 151–158. [CrossRef]
23. Devos, K.M.; Bryan, G.J.; Collins, A.J.; Stephenson, P.; Gale, M.D. Application of two microsatellite sequences in wheat storage proteins as molecular markers. *Theory Appl. Genet.* **1995**, *90*, 247–252. [CrossRef] [PubMed]
24. Somers, D.J.; Zhou, Z.; Bebeli, P.J.; Gustafson, J.P. Repetitive, genome-specific probes in wheat (*Triticum aestivum* L. em Thell) amplified with minisatellite core sequences. *Theory Appl. Genet.* **1996**, *93*, 982–989. [CrossRef] [PubMed]
25. International Wheat Genome Sequencing Consortium (IWGSC). Shifting the limits in wheat research and breeding using a fully annotated reference genome. *Science* **2018**, *361*, eaar7191. [CrossRef] [PubMed]
26. Ling, H.Q.; Ma, B.; Shi, X.; Liu, H.; Dong, L.; Sun, H.; Cao, Y.; Gao, Q.; Zheng, S.; Li, Y.; et al. Genome sequence of the progenitor of wheat a subgenome *Triticum urartu*. *Nature* **2018**, *557*, 424–428. [CrossRef] [PubMed]
27. Luo, M.C.; Gu, Y.Q.; Puiu, D.; Wang, H.; Twardziok, S.O.; Deal, K.R.; Huo, N.; Zhu, T.; Wang, L.; Wang, Y.; et al. Genome sequence of the progenitor of the wheat D genome *Aegilops tauschii*. *Nature* **2017**, *551*, 498–502. [CrossRef] [PubMed]
28. Avni, R.; Nave, M.; Barad, O.; Baruch, K.; Twardziok, S.O.; Gundlach, H.; Hale, I.; Mascher, M.; Spannagl, M.; Wiebe, K.; et al. Wild emmer genome architecture and diversity elucidate wheat evolution and domestication. *Science* **2017**, *357*, 93–97. [CrossRef]
29. Lang, T.; Li, G.R.; Wang, H.J.; Yu, Z.H.; Chen, Q.H.; Yang, E.N.; Fu, S.L.; Tang, Z.X.; Yang, Z.J. Physical location of tandem repeats in the wheat genome and application for chromosome identification. *Planta* **2018**. [CrossRef]
30. Evtushenko, E.V.; Levitsky, V.G.; Elisafenko, E.A.; Gunbin, K.V.; Belousov, A.I.; Šafář, J.; Doležel, J.; Vershinin, A.V. The expansion of heterochromatin blocks in rye reflects the co-amplification of tandem repeats and adjacent transposable elements. *BMC Genom.* **2016**, *17*, 337. [CrossRef]
31. Zeng, Z.X.; Hu, L.J.; Li, G.R.; Liu, C.; Yang, Z.J. Phenotypic and epigenetic changes occurred during the autopolyploidization of *Aegilops tauschii*. *Cereal Res. Commun.* **2012**, *49*, 476–485. [CrossRef]
32. Yang, Z.J.; Li, G.R.; Feng, J.; Jiang, H.R.; Ren, Z.L. Molecular cytogenetic characterization and disease resistance observation of wheat—*Dasypyrum breviaristatum* partial amphiploid and its derivatives. *Hereditas* **2005**, *142*, 80–85. [CrossRef] [PubMed]
33. Benson, G. Tandem repeats finder: A program to analyze DNA sequences. *Nucleic Acids Res.* **1999**, *27*, 573–580. [CrossRef]
34. Li, W.Z.; Godzik, A. Cd-hit: A fast program for clustering and comparing large sets of protein or nucleotide sequences. *Bioinformatics* **2006**, *22*, 1658–1659. [CrossRef] [PubMed]
35. Kato, A.; Lamb, J.C.; Birchler, J.A. Chromosome painting using repetitive DNA sequences as probes for somatic chromosome identification in maize. *Proc. Natl. Acad. Sci. USA* **2004**, *101*, 13554–13559. [CrossRef] [PubMed]
36. Tang, Z.X.; Yang, Z.J.; Fu, S.L. Oligonucleotides replacing the roles of repetitive sequences pAs1, pSc119.2, pTa-535, pTa71, CCS1, and pAWRC.1 for FISH analysis. *J. Appl. Genet.* **2014**, *55*, 313–318. [CrossRef] [PubMed]
37. Fu, S.L.; Chen, L.; Wang, Y.Y.; Li, M.; Yang, Z.J.; Qiu, L.; Yan, B.J.; Ren, Z.L.; Tang, Z.X. Oligonucleotide probes for ND-FISH analysis to identify rye and wheat chromosomes. *Sci. Rep.* **2015**, *5*, 10552. [CrossRef] [PubMed]
38. Rozas, J.; Ferrer-Mata, A.; Sánchez-DelBarrio, J.C.; Guirao-Rico, S.; Librado, P.; Ramos-Onsins, S.E.; Sánchez-Gracia, A. DnaSP 6: DNA sequence polymorphism analysis of large data sets. *Mol. Biol. Evol.* **2017**, *34*, 3299–3302. [CrossRef]
39. Larkin, M.A.; Blackshields, G.; Brown, N.P.; Chenna, R.; McGettigan, P.A.; McWilliam, H.; Valentin, F.; Wallace, I.M.; Wilm, A.; Lopez, R.; et al. Clustal W and Clustal X version 2.0. *Bioinformatics* **2007**, *23*, 2947–2948. [CrossRef] [PubMed]

40. Kumar, S.; Stecher, G.; Tamura, K. MEGA7: Molecular evolutionary genetics analysis version 7.0 for bigger datasets. *Mol. Biol. Evol.* **2016**, *33*, 1870–1874. [CrossRef] [PubMed]
41. Schneider, T.D.; Stephens, R.M. Sequence logos: A new way to display consensus sequences. *Nucleic Acids Res.* **1990**, *18*, 6097–6100. [CrossRef] [PubMed]
42. Crooks, G.E.; Hon, G.; Chandonia, J.M.; Brenner, S.E. WebLogo: A sequence logo generator. *Genome Res.* **2004**, *14*, 1188–1190. [CrossRef] [PubMed]
43. Li, G.R.; Gao, D.; Zhang, H.J.; Li, J.B.; Wang, H.J.; La, S.X.; Ma, J.W.; Yang, Z.J. Molecular cytogenetic characterization of *Dasypyrum breviaristatum* chromosomes in wheat background revealing the genomic divergence between *Dasypyrum* species. *Mol. Cytogenet.* **2016**, *9*, 6. [CrossRef] [PubMed]
44. Li, J.B.; Lang, T.; Li, B.; Yu, Z.H.; Wang, H.J.; Li, G.R.; Yang, E.N.; Yang, Z.J. Introduction of *Thinopyrum intermedium* ssp. *trichophorum chromosomes to wheat by trigeneric hybridization involving Triticum, Secale and Thinopyrum genera. Planta* **2017**, *245*, 1121–1135. [CrossRef] [PubMed]
45. Wicker, T.; Gundlach, H.; Spannagl, M.; Uauy, C.; Borrill, P.; Ramírez-González, R.H.; De Oliveira, R.; International Wheat Genome Sequencing Consortium (IWGSC); Mayer, K.F.X.; Paux, E.; et al. Impact of transposable elements on genome structure and evolution in bread wheat. *Genome Biol.* **2018**, *19*, 103. [CrossRef] [PubMed]
46. Martienssen, R.A.; Baulcombe, D.C. An unusual wheat insertion sequence (WIS1) lies upstream of an α-amylase gene in hexaploid wheat, and carries a "minisatellite" array. *Mol. Gen. Genet.* **1989**, *217*, 401–410. [CrossRef] [PubMed]
47. Tang, S.Y.; Tang, Z.X.; Qiu, L.; Yang, Z.J.; Li, G.R.; Lang, T.; Zhu, W.Q.; Zhang, J.H.; Fu, S.L. Developing new Oligo probes to distinguish specific chromosomal segments and the A, B, D genomes of wheat (*Triticum aestivum* L.) using ND-FISH. *Front. Plant Sci.* **2018**, *9*, 1104. [CrossRef] [PubMed]
48. Mascher, M.; Gundlach, H.; Himmelbach, A.; Beier, S.; Twardziok, S.O.; Wicker, T.; Radchuk, V.; Dockter, C.; Hedley, P.E.; Russell, J.; et al. A chromosome conformation capture ordered sequence of the barley genome. *Nature* **2017**, *544*, 427–433. [CrossRef] [PubMed]
49. Daccord, N.; Celton, J.M.; Linsmith, G.; Becker, C.; Choisne, N.; Schijlen, E.; van de Geest, H.; Bianco, L.; Micheletti, D.; Velasco, R.; et al. High-quality de novo assembly of the apple genome and methylome dynamics of early fruit development. *Nat. Genet.* **2017**, *49*, 1099–1106. [CrossRef] [PubMed]
50. Jiao, Y.; Peluso, P.; Shi, J.; Liang, T.; Stitzer, M.C.; Wang, B.; Campbell, M.S.; Stein, J.C.; Wei, X.; Chin, C.S.; et al. Improved maize reference genome with single-molecule technologies. *Nature* **2017**, *546*, 524–527. [CrossRef] [PubMed]
51. Kuhn, G.C.S.; Heslop-Harrison, J.S. Characterization and genomic organization of PERI, a repetitive DNA in the *Drosophila buzzatii* cluster related to DINE-1 transposable elements and highly abundant in the sex chromosomes. *Cytogenet. Genome Res.* **2011**, *132*, 79–88. [CrossRef] [PubMed]
52. Wei, K.H.C.; Grenier, J.K.; Barbash, D.A.; Clark, A.G. Correlated variation and population differentiation in satellite DNA abundance among lines of Drosophila melanogaster. *Proc. Natl. Acad. Sci. USA* **2014**, *111*, 18793–18798. [CrossRef]
53. Macas, J.; Navrátilová, A.; Koblížková, A. Sequence homogenization and chromosomal localization of VicTR-B satellites differ between closely related Vicia species. *Chromosoma* **2006**, *115*, 437–447. [CrossRef] [PubMed]
54. Kuhn, G.C.S.; Sene, F.M.; Moreira-Filho, O.; Schwarzacher, T.; Heslop-Harrison, J.S. Sequence analysis, chromosomal distribution and long-range organization show that rapid turnover of new and old pBuM satellite DNA repeats leads to different patterns of variation in seven species of the *Drosophila buzzatii* cluster. *Chromosome Res.* **2008**, *16*, 307–324. [CrossRef] [PubMed]
55. Martinsen, L.; Venanzetti, F.; Johnsen, A.; Sbordoni, V.; Bachmann, L. Molecular evolution of the pDo500 satellite DNA family in *Dolichopoda* cave crickets (*Rhaphidophoridae*). *BMC Evol. Biol.* **2009**, *9*, 301. [CrossRef] [PubMed]
56. Plohl, M.; Petrović, V.; Luchetti, A.; Ricci, A.; Šatović, E.; Passamonti, M.; Mantovani, B. Long-term conservation vs high sequence divergence: The case of an extraordinarily old satellite DNA in bivalve mollusks. *Heredity* **2010**, *104*, 543–551. [CrossRef] [PubMed]
57. Li, G.R.; Lang, T.; Yang, E.N.; Liu, C.; Yang, Z.J. Characterization and phylogenetic analysis of α-gliadin gene sequences reveals significant genomic divergence in Triticeae species. *J. Genet.* **2014**, *93*, 725–731. [CrossRef] [PubMed]

58. International Wheat Genome Sequencing Consortium (IWGSC). A chromosome-based draft sequence of the hexaploid bread wheat (*Triticum aestivum*) genome. *Science* **2014**, *345*, 1251788. [CrossRef]
59. Badaeva, E.D.; Friebe, B.; Gill, B.S. Genome differentiation in *Aegilops*. 1. Distribution of highly repetitive DNA sequences on chromosomes of diploid species. *Genome* **1996**, *39*, 293–306. [CrossRef]
60. Ruban, A.S.; Badaeva, E.D. Evolution of the S-Genomes in *Triticum-Aegilops* alliance: Evidences from chromosome analysis. *Front. Plant Sci.* **2018**, *9*, 1756. [CrossRef]
61. Luchetti, A.; Cesari, M.; Carrara, G.; Cavicchi, S.; Passamonti, M.; Scali, V.; Mantovani, B. Unisexuality and molecular drive: Bag320 sequence diversity in Bacillus taxa (*Insecta Phasmatodea*). *J. Mol. Evol.* **2003**, *56*, 587–596. [CrossRef] [PubMed]
62. Robles, F.; de la Herrán, R.; Ludwig, A.; Ruiz Rejón, C.; Ruiz Rejón, M.; Garrido-Ramos, M.A. Evolution of ancient satellite DNAs in sturgeon genomes. *Gene* **2004**, *338*, 133–142. [CrossRef] [PubMed]
63. Meštrović, N.; Castagnone-Sereno, P.; Plohl, M. Interplay of selective pressure and stochastic events directs evolution of the MEL172 satellite DNA library in root-knot nematodes. *Mol. Biol. Evol.* **2006**, *23*, 2316–2325. [CrossRef] [PubMed]
64. Navajas-Pérez, R.; Quesada del Bosque, M.E.; Garrido-Ramos, M.A. Effect of location, organization, and repeat-copy number in satellite-DNA evolution. *Mol. Genet. Genom.* **2009**, *282*, 2316–2325. [CrossRef] [PubMed]
65. Giovannotti, M.; Cerioni, P.N.; Splendiani, A.; Ruggeri, P.; Olmo, E.; Barucchi, V.C. Slow evolving satellite DNAs: The case of a centromeric satellite in *Chalcides ocellatus* (Forskål, 1775) (*Reptilia, Scincidae*). *Amphibia-Reptilia* **2013**, *34*, 401–411. [CrossRef]
66. Komuro, S.; Endo, R.; Shikata, K.; Kato, A. Genomic and chromosomal distribution patterns of various repeated DNA sequences in wheat revealed by a fluorescence in situ hybridization procedure. *Genome* **2013**, *56*, 131–137. [CrossRef]
67. Danilova, T.V.; Akhunova, A.R.; Akhunov, E.D.; Friebe, B.; Gill, B.S. Major structural genomic alterations can be associated with hybrid speciation in *Aegilops markgrafii* (*Triticeae*). *Plant J.* **2017**, *92*, 317–330. [CrossRef]
68. Lang, T.; La, S.X.; Li, B.; Yu, Z.H.; Chen, Q.H.; Li, J.B.; Yang, E.N.; Li, G.R.; Yang, Z.J. Precise identification of wheat—*Thinopyrum intermedium* translocation chromosomes carrying resistance to wheat stripe rust in line Z4 and its derived progenies. *Genome* **2018**, *61*, 177–185. [CrossRef]
69. Yu, Z.H.; Wang, H.J.; Xu, Y.F.; Li, Y.S.; Lang, T.; Yang, Z.J.; Li, G.R. Characterization of chromosomal rearrangement in new wheat—*Thinopyrum intermedium* addition lines carrying *Thinopyrum*—Specific grain hardness genes. *Agronomy* **2019**, *9*, 18. [CrossRef]

Review

The Effect of Chromosome Structure upon Meiotic Homologous and Homoeologous Recombinations in Triticeae

Tomás Naranjo

Departamento de Genética, Fisiología y Microbiología, Facultad de Biología, Universidad Complutense de Madrid, 28040 Madrid, Spain; toranjo@bio.ucm.es; Tel.: +34-913945040

Received: 23 August 2019; Accepted: 11 September 2019; Published: 14 September 2019

Abstract: The tribe *Triticeae* contains about 500 diploid and polyploid taxa, among which are important crops, such as wheat, barley and rye. The phylogenetic relationships, genome compo-sition and chromosomal architecture, were already reported in the pioneer genetic studies on these species, given their implications in breeding-related programs. Hexaploid wheat, driven by its high capacity to develop cytogenetic stocks, has always been at the forefront of these studies. Cytogenetic stocks have been widely used in the identification of homoeologous relationships between the chromosomes of wheat and related species, which has provided valuable information on genome evolution with implications in the transfer of useful agronomical traits into crops. Meiotic recombination is non-randomly distributed in the *Triticeae* species, and crossovers are formed in the distal half of the chromosomes. Also of interest for crops improvement is the possibility of being able to modulate the intraspecific and interspecific recombination landscape to increase its frequency in crossover-poor regions. Structural changes may help in this task. In fact, chromosome truncation increases the recombination frequency in the adjacent intercalary region. However, structural changes also have a negative effect upon recombination. Gross chromosome rearrangements produced in the evolution usually suppress meiotic recombination between non-syntenic homoeologs. Thus, the chromosome structural organization of related genomes is of great interest in designing strategies of the introgression of useful genes into crops.

Keywords: chromosome rearrangements; meiotic recombination; crossover distribution; *Triticeae*; wheat; barley; rye

1. Introduction

The tribe *Triticeae Dumort* belongs to the subfamily *Pooideae* of the grass family *Poaceae* (*Gramineae*). About 500 taxa are included in this tribe. The number and designation of genera, species and subspecies depend upon the taxonomic criterion [1,2]. All species are considered to have a monophyletic origin, that is, they have a common ancestor, which is theoretically estimated to diverge from the other *Poaceae* 25 million years ago (MYA) [3,4]. Among the about 360 species in the tribe, 80 are diploids, the majority are allopolyploids derived from interspecific or intergeneric hybrids, most likely with progenitor species repeatedly involved in different hybridization events and increasing polyploidy levels. Only a few species are considered autopolyploids, which originated mainly from intraspecific hybridization followed by chromosome doubling [5]. All species have a basic haploid chromosome number of $x = 7$; diploid species have $2n = 14$ chromosomes, and polyploids possess multiples of 14 chromosomes, the tetraploids $2n = 4x = 28$, and the hexaploids $2n = 6x = 42$. *Elymus* shows the most complex series and the highest level ($12x$) of ploidy [2].

Some species are self-pollinated, and others are cross-pollinated. About 75% of *Triticeae* species are perennials that grow in Temperate and Artic regions, mainly in the Northern Hemisphere, while the annuals spread through the Eastern Mediterranean and Central Asiatic regions [2].

The tribe is the subject of many research areas such as archaeology, genetics, plant breeding and evolutionary biology, but the main interest is due to its global economic significance, as it includes grain crops such as wheat (*Triticum*), barley (*Hordeum*) and rye (*Secale*), as well as a considerable number of grasses used for animal feed or rangeland protection. Given the enormous capacity for interspecific hybridization among most of the *Triticeae* taxa, wild species represent an accessible source of genetic variability that can be used to improve crops. The taxonomy of the tribe is complex, and has varied over time according to the taxonomic criteria and methodologies used [1]. Probably, interspecific hybridization and a large number of allopolyploid species are the key factors in the appearance of inconsistencies between different phylogenetic trees. However, the role of the wild species as potential donors of useful agronomic traits to be introgressed into their cultivated relatives, represents an additional incentive in resolving the supposed evolutionary relationships among them.

With regards to the divergence of different genera there is a common consent that *Psathyrostachys* and *Hordeum* diverged early on from the rest of the *Triticeae*. *Aegilops* and *Triticum* are closely related genera with a more recent origin. The estimates of the divergent times suggest that *Hordeum* diverged 11 MYA, *Secale* 7 MYA and the *Aegilops-Triticum* complex 2.5–4.5 MYA [4]. *Hordeum* originated in western Eurasia, and later extended over The Northern Hemisphere, South Africa and South America. It contains about 33 species. Four genomes have been described H, I, Xa and Xu, which are present in diploid (2*x*), tetraploid (4*x*) and hexaploid (6*x*) taxa [6]. The H genome is present in barley, *H. vulgare*, and *H. bulbosum*, Xu in *H. murinum*, Xa in *H. marinum*, and the remaining species share variants of the I genome. Barley, used to feed livestock, and malted for beer or whisky (also whiskey) production, is the economically most important species in the genus, with a world production of about 147.40 million metric tons in 2017 (http://www.fao.org/faostat/es/#data/QC). It ranks the fourth in both quantity produced and in area of cultivation of cereal crops in the world. The barley genome has a size of 5.1 Gb, and contains 30,400 genes [7]. Its chromosomes are designated 1H to 7H according to the homoeologous relationships relative to the wheat chromosomes. The taxonomy of the genus *Secale* has been a matter of disagreement. The about 15 species initially recognized were reduced to three, *S. strictum*, *S. sylvestre* and *S. cereale*. This *S. strictum*, the perennial species of the genus, is considered the ancestral taxon. The other two species are annual, and *S. sylvestre* diverged first from *S. strictum*, while *S. cereale* is the youngest species of the group, and includes also *S. vavilovi* [8]. Rye (*S. cereale*) is the sixth most important cereal in production with a total of 13.73 million metric tons in 2017 (http://www.fao.org/faostat/es/#data/QC). Rye also has a large genome (8.1 Gb, nearly 50% larger than the barley genome) with 31,000 detected genes [9]. Chromosomes of rye are designated 1R to 7R, according to the homoeologous relationships relative to wheat chromosomes.

The wheat-*Aegilops* complex consists of three genera, *Amblyopirum*, *Aegilops* and *Triticum*, with one, 10 and two diploid species each, respectively. Based on the genome divergence, these diploid species were classified in seven groups: T-group (*A. muticum*); S-group (*Ae. speltoides*, *Ae. bicornis*, *Ae. longissima*, *Ae. sharonensis*, and *Ae. searsii*); A-group (*T. monococcum* and *T. urartu*); D-group (*Ae. tauschii*); C-group (*Ae. caudata*); M-group (*Ae. comosa* and *Ae. uniaristata*); U-group (*Ae. umbellulata*) [2]. In addition to the diploids, the *Aegilops* genus contains 10 allotetraploid species: *Ae. biuncialis* (UUMM), *Ae. geniculata* (MMUU), *Ae. neglecta* (UUMM), *Ae. columnaris* (UUMM), *Ae triuncialis* (UUCC), *Ae. kotschyi* (SSUU), *Ae. peregrina* (SSUU), *Ae. cylindrica* (CCDD), *Ae. crassa* (DDMM), and *Ae. ventricosa* (DDNN), and four allohexaploids: *Ae. recta* (UUMMNN), *Ae. vavilovi* (DDMMSS), *Ae. crassa* (DDDDMM), and *Ae. juvenalis* (DDMMUU).

The genus *Triticum* contains two tetraploid species, *T. turgidum* (AABB) and *T. timopheevii* (AAGG), the allohexaploid *T. aestivum* (AABBDD) and the autoallohexaploid *T. zhukovskyi* (AAGGA^{m}A^m). *T. turgidum* and *T. aestivum* constitute the Emmer lineage, and *T. timopheevii* and *T. zhukovskyi* constitute the Timopheevii lineage of polyploid wheats. *Triticum* is the most important genus of the tribe *Triticeae* for its agricultural value. Its ranks the second after corn in grain production, contributing about 20% of the total calories consumed by humans.

The world production of wheat reached 771.72 million metric tons in 2017 (http://www.fao.org/faostat/es/#data/QC), most of which is bread wheat, *T. aestivum*. Several studies on the divergence time of the diploid genomes of the wheat-*Aegilops* complex agree in showing the A and S genomes as the basal branches of the group, while the D genome diverged later [10–12]. Feldman and Levy [2] suggested a reticulate evolution in the group, in which the primitive diploid species diverged theoretically 2.5–3.0 MYA, and the more recent ones 2.5 MYA. The three ploidy levels of the wheat group contain cultivated species. One diploid species, *T. monococcum*, and the two tetraploids, were domesticated with the birth of agriculture about 10,000 years ago, while *T. aestivum* and *T. zhukovskyi* are considered to be originated under the cultivation of *T. turgidum* and *T. timopheevii*, respectively. Wild tetraploid wheat, *T. turgidum* ssp. *dicoccoides* (AABB) is believed to have originated within the past few hundred thousand years [4] from the hybridization of wild diploid einkorn wheat, *T. urartu* (AA), with a close relative of *Ae. speltoides* (SS, where S is closely related to B) [13–15]. The hexaploid *T. aestivum* (AABBDD) originated from a second hybridization event between *T. turgidum* and the wild diploid progenitor of the D genome, *Ae. tauschii* [16,17]. *T. timopheevii* originated 0.4 MYA [4,12], its A-genome was derived from *T. urartu* [13], while the G-genome originated from the S-genome of *Ae. speltoides* [15,18–23]. The hexaploid wheat *T. zhukovskyi* originated from the hybridization of *T. timopheevii* and *T. monococcum* [13,24]. Bread wheat has a large genome (17 Gb), whose reference sequence has been identified with an estimated coverage of 94%, giving access to 107,891 high-confidence genes [25,26]. A 10.5-Gb genome assembly composed of 14 pseudomolecule sequences representing the 14 chromosomes of wild emmer wheat (genome size estimated to be 12 Gb) was also reported [27]. Other wheat-relative genome sequences, including those of *T. urartu* (4.94 Gb) and *Ae. tauschii* (4.5 Gb), have been published [28,29].

Many studies carried out in order to identify the diploid progenitors of the allopolyploid species in the wheat-*Aegilops* complex were based in the genome analysis developed by Kihara [30–33]. This method assumed that the genetic architecture of the different genomes present in a given allopolyploid is preserved as in their diploid progenitors, making possible their identification through the analysis of meiotic pairing in interspecific crosses. A high frequency of bivalent pairing at metaphase I between the chromosomes of the diploid and polyploid species identified the homology of such genomes. Meiotic pairing also has implications on the possibility of transferring genes of agronomic interest from wild species to cultivated ones, especially to common wheat. Homologous recombination can be used in the transfer of genes from species of the primary and secondary gene pool, which share homologous genomes with common wheat. After direct hybridization and homologous recombination, useful genes can be selected in the offspring of the hybrids and in subsequent backcrosses [34]. In the case of species belonging to the tertiary gene pool, genomes are more differentiated relative to those of wheat, and any gene transfer to be achieved needs the occurrence of a recombination between homoeologous chromosomes. This became effective after the discovery of the *Ph1* gene that suppresses recombination between homoeologous chromosomes in wheat [35–39]. In the absence of *Ph1*, recombination is possible between the homoeologous chromosomes of wheat or between those of wheat and other species. Alternatively, *Ph* suppressors present in *Ae. speltoides* can be used to induce homoeologous recombination. Different genes have been transferred from the *Aegilops* species via such homoeologous recombination [40]. On the other hand, meiotic recombination appears also to be the foci of current research projects because of the importance of extending their occurrence to all genomic regions, especially to those that are usually restricted. In this review, I will summarize all of the features of meiotic recombination with implications on breeding-related programs, especially those concerning polyploid wheats.

2. Chromosome Structure in Triticeae

All species of the tribe *Triticeae* are assumed to have evolved from a common ancestor. In the species divergence, chromosomes could adopt different evolutionary paths, which can be established after unraveling the content and order of the genes present in the homoeologous linkage groups. In this task, bread wheat has played a central role.

Hexaploid wheat, *T. aestivum*, is the first allopolyploid in which all 21 chromosomes were identified according to its genomes of origin (A, B or D) and homoeologous groups. Such a chromosome classification was later used as a reference in the identification of the homoeologous relationships of chromosomes from other genomes of the tribe. Aneuploid sets of monosomics, nullisomis, telocentrics, trisomics and tetrasomics, as well as the 42 nullisomic-tetrasomic combinations obtained by Sears in cv. Chinese Spring [41,42], were indispensable materials used for this purpose. Unpaired chromosomes in the crosses of 21 monosomics of bread wheat and tetraploid wheat AABB were allocated to the D genome [41]. Chromosomes of the A and B genomes were identified in crosses between 13 ditelocentric lines and a synthetic tetraploid AADD [43]. Six telocentrics that formed a heteromorphic bivalent with a normal chromosome in the AABDD hybrids were assigned to the A genome, and the seven telocentrics that appeared as univalents were assigned to the B genome. The remaining chromosome, which belonged to group 4 [41], was determined to be 4A. Further studies demonstrated the initial misclassification of the chromosomes 4A and 4B, and they were reassigned as 4B and 4A, respectively [44].

Its hexaploid constitution confers upon bread wheat the ability of tolerating the loss of complete chromosomes, as well as the addition of chromosomes from related species without drastically affecting the viability of the plant. This ability and the crossability of wheat with different species of genera *Aegilops, Haynaldia, Secale, Thinopyrum, Elymus, Leymus, Elytrigia, Roegneria* or *Hordeum* have permitted to develop wheat-alien addition and substitution lines. Sets of wheat-alien additions are derived from backcrosses of interspecific hybrids, or synthetic allopolyploids, with wheat. Such lines contain one chromosome (monosomic addition) or one chromosome pair (disomic addition) from a given specie added to the entire chromosome complement of common wheat. Additions from more than 20 related species have been reported in bread wheat [45]. Once a complete set of wheat-alien additions is obtained, the added chromosomes are assigned to homoeologous groups based on their genetic affinity to wheat chromosomes. In wheat-alien substitution lines a pair of alien homologues are substituted for a pair of wheat chromosomes. These lines are usually more stable than addition lines, but their production requires the use of wheat-alien additions. The standard method to obtain alien substitutions is to cross a disomic addition with a wheat monosomic of the same homoeologous group of the alien chromosome, and to select the disomic substitution in the offspring of the double monosomic obtained in the cross. In the case where the homoeologous relationships of the addition lines were not established, it is possible to cross a wheat-alien addition with seven monosomics. Among the substitutions obtained, only one involving homoeologous chromosomes will produce genetic compensating and fertile plants, thus identifying the homoeologous group of the alien chromosome. The ability of an alien chromosome to compensate for the absence of a wheat chromosome resembles the nullisomic-tetrasomic compensation test used by Sears [42] in the identification of the homoeologous chromosomes of wheat, and is the most efficient method of identifying the homoeology of alien chromosomes to wheat. Telocentrics produced through a mis-division of the added chromosomes have also been recovered, and these increase the richness of cytogenetic stocks available in wheat.

Once the homoeologous relationships of chromosomes from different genomes are identified, the level of synteny that they share can be assessed using different comparative analysis. Meiotic pairing [46], physical location of low copy DNA probes [47], genetic maps [48,49] and genome sequencing [50] are different approaches used to study whether collinearity of homoeologous chromosomes is preserved or broken down by rearrangements produced during genome differentiation.

2.1. Durum and Bread Wheats

Meiotic pairings have been used to study the chromosome structure of bread wheat genomes [46]. The wheat chromosome arms that paired at metaphase I in wheat × rye hybrids lacking either the *Ph1* or *Ph2* were identified by C-banding. In most cases associations occurred between homoeologous arms, but some associations were produced between chromosome arms of different homoeologous groups.

The 4AL arm was paired with 7AS and 7DS, 5AL with 4BL and 4DL, and 7BS with 5BL and 5DL. In addition, infrequent associations of 4AL-4DS, 4AS-4BL and 4AS-4DL were also observed. Such results indicated normal homoeology for all wheat chromosome arms except 4AS, 4AL, 5AL and 7BS. The arms 4AL, 5AL and 7BS are involved in a cyclic translocation and chromosome 4A in a pericentric inversion produced in the evolution of wheat. Meiotic pairing between chromosomes of tetraploid wheat *T. turgidum* and *T. aestivum* suggested the following sequence of chromosome rearrangements [51]. The arms 4AL and 5AL were involved in a reciprocal translocation, T(4AL-5AL)1 [52], produced during the evolution of the A genome, which was transmitted from *T. urartu* to *T. turgidum*. This and other chromosome rearrangements are described according to the standard nomenclature [53]. A second reciprocal translocation, T(4AL-7BS)2, and the pericentric inversion of chromosome 4A, Inv(4AS.4AL)1, occurred at the tetraploid stage. All of these chromosome rearrangements were transmitted to *T. aestivum*.

Differentiation of the chromosome structure of wheat genomes was also addressed by physical mapping. More than 60 full-length cDNAs were used as fluorescence in situ hybridization (FISH) probes, whose position was identified in the arms of the 21 chromosomes of hexaploid wheat [47]. Markers were selected to produce three different signals per chromosome arm located in proximal, intercalary and distal positions, respectively. Most of these probes hybridized to all of the three wheat homoeologs. The order and relative positions of the markers was similar in all chromosome arms, except in chromosome arms 2AS, 4AS, 4AL, 4BL, 6AS and 7BS. Such exceptions are the result of the evolutionary rearrangements of 4A, 5AL and 7BS, as indicated above, or the chromosome rearrangements which appeared in Chinese Spring and other cultivars.

A genetic map of wheat chromosomes, which was based on the arm location of 800 restriction fragment length polymorphism (RFLP) markers using aneuploidy stocks, confirmed the chromosome rearrangements detected with chromosome pairing [48]. Construction of linkage maps involving homoeologous group 4, 5 and 7 chromosomes revealed the presence of a paracentric inversion on chromosome arm 4AL, Inv(4AL.4AL)1 [49,54]. The use of the deletion stocks of wheat [55] allowed us to establish the distribution of different RFLPs markers along the wheat group 4 chromosomes, and a more precise identification of rearranged segments on chromosome 4A, as well as a pericentric inversion on chromosome 4B [56]. Using nulli-tetrasomic and ditelosomic lines and the deletion set of wheat group 4 chromosomes, the distribution of 1,918 expressed sequence tag (EST) loci along different bins of chromosomes 4A, 4B and 4D, was reported [57]. In addition to the identification of inversions Inv(4AS.4AL) and Inv(4AL.4AL)1 and the translocated segments from 5AL and 7BS, another pericentric inversion, Inv(4AS.4AL)2, was detected on chromosome 4A. Comparison of genome sequences of *Ae. tauschii* and wild *T. turgidum* ssp. *dicoccoides* was employed to identify the breakpoints in the rearrangements of chromosome 4A, 5A and 7B [52]. The breakpoints harboring the pericentric inversion Inv(4AS.4AL)1 were located in sequences containing satellite DNA, while those of reciprocal translocations T(4AL-5AL) and T(4AL-7BS) were situated in non-repeated DNA sequences. The breakpoints of the paracentric inversion Inv(4AL.4AL)1 overlapped with the breakpoints of Inv(4AS.4AL)1 and translocation T(4AL-7BS), which suggested that these three rearrangements occurred simultaneously. The breakpoints of pericentric inversion Inv(4AS.4AL)2 were not identified.

The shotgun sequence obtained from individual chromosome arms of Chinese Spring provided a map of 551 homoeologous genes, which denoted the presence of pericentric inversions in at least 10 of the 21 chromosomes [58]. It is likely that some of such rearrangements occurred during the production of the aneuploid lines used in the study. A mapping population of 445 recombinant inbred lines (RILs) obtained from the cross of durum wheat cv. 'Langdon' x wild emmer was developed and genotyped with single nucleotide polymorphism (SNP) markers to construct a high-density map of 2,650 segregating markers. An alignment of SNP markers in *T. turgidum* compared to that in *Brachypodium distachyon*, rice and sorghum, revealed the presence of 15 structural chromosome rearrangements, in addition to the already known rearrangements [59].

Intra-chromosomal translocations of short segments, a reciprocal translocation between 3B and 6B, a non-reciprocal translocation of a short 6BS segment to 7BL, and three large and one small paracentric inversions, were discovered. Genetic diversity in the rearranged chromosome 4A showed that the rearrangements could occur in the primitive tetraploid wheat.

Minor chromosome rearrangements due to gene locus deletion or gene locus duplication were detected in a sample of 3,159 Chinese Spring A- and D-genome gene loci. Such microsynteny perturbations occurred at both the diploid and polyploid levels, and showed a non-random distribution. Their occurrence correlated positively with the distance of the locus from the centromere [10].

Comparative studies revealed significant macro-collinearity between cereal genomes supporting that rice, sorghum and *Triticeae* originated from a common ancestor with a basic number of x = 12 chromosomes [60]. Rice maintains this chromosome number, but a reduction to x = 10 and x = 7 occurred in the divergence of sorghum and the ancestor of *Triticeae*, respectively. Comparison of the genome of *Ae. tauschii* with the rice and sorghum genomes revealed different chromosome rearrangements produced in the evolution of sorghum and *Triticeae*, and that reduction of the chromosome number was mainly produced by insertion of an entire chromosome in a break produced in the pericentromeric region of another chromosome. In some instance, the inserted chromosome underwent telomeres fusion accompanied with a break in the pericentromeric region [61]. Of 50 rearrangements detected, 40 were assigned to the *Ae. tauschii* lineage, while two and eight occurred in the evolution of rice and sorghum, respectively. This result suggested a more accelerated evolution of the large genome of the *Triticeae* species.

2.2. *T. timopheevii*

Sequential N-banding and in situ hybridization revealed two intergenomic translocations present in *T. timopheevii* [62]. These and other translocations were evidenced from chromosome pairing analysis in the hybrids between *T. turgidum* and *T. timopheevii* (AA^{t}BG), between *T. aestivum* and *T. timopheevii* (AA^{t}BGD) and between *T. aestivum* and *T. turgidum* (AABBD) [63,64]. Chromosomes 1A^t, 2A^t, 5A^t, 7A^t, 2G, 3G, 5G and 6G do not structurally differ from their counterparts of the A and B genomes. The remaining chromosomes are involved in five reciprocal translocations: T(4A^{t}L-5A^{t}L), which was inherited from *T. urartu*, and translocations T(6A^{t}S-1GS), T(1GS-4GS), T(4GS-4A^{t}L) and T(4A^{t}L-3A^{t}L), which were produced at the tetraploid stage, probably in that sequence. While A-A and A-A^t chromosome associations showed similar frequencies in interspecific hybrids, B-B associations were more frequent than B-G associations, denoting a higher differentiation of the B and G genomes relative to the A and A^t genomes. This behavior in addition to differences in the chromosome structure between *T. turgidum* and *T. timopheevii* support a diphyletic origin of these two tetraploid species. Comparative mapping using microsatellite markers confirmed the presence of translocation T(6A^{t}S-1GS) [65,66], and revealed a paracentric inversion on 6A^{t}L [65]. Genetic maps with a higher density are required for a more reliable knowledge of the structure of the A^t and G genomes.

2.3. The Sitopsis Section of the Genus Aegilops

Different studies have been reported on the chromosome structure of *Ae. longissima* (S^{l}S^l), *Ae. sharonensis* (S^{sh}S^{sh}), and *Ae. speltoides* (SS), but little is known in the other two species of the *Sitopsis* section, *Ae. bicornis* and *Ae. searsii*. Chromosome pairing analysis in hybrids of hexaploid wheat and *Ae. longissima* showed that the arms 4S^{l}L and 7S^{l}L are involved in an evolutionary translocation, while 4S^{l}S, 7S^{l}S and chromosome arms 1S^l, 2S^l, 3S^l, 5S^l and 6S^l show normal homoeology to wheat [67,68]. Chromosomes of *Ae. sharonensis* and *Ae. speltoides* and their arms show normal homoeologous relationships to wheat as based on the wheat-*Aegilops* chromosome pairing [69,70].

A comparative map of 67 RFLP loci constructed in *Ae. longissima* confirmed that chromosomes $1S^l$, $2S^l$, $3S^l$, $5S^l$ and $6S^l$, the arms $4S^lS$ and $7S^lS$, and the proximal regions of $4S^lL$ and $7S^lL$ are collinear with wheat D-genome chromosomes, and that the distal part of $4S^lL$ was translocated from $7S^lL$ [71].

A linkage map of *Ae. sharonensis* based on 377 Diversity Array Technology (DArT) and 12 simple sequence repeat (SSR) markers comprised 10 linkage groups, four of which were collinear with wheat chromosomes, and assigned to chromosomes $1S^{sh}$, $2S^{sh}$, $3S^{sh}$ and $6S^{sh}$ [72]. A further linkage map constructed with 727 oligo pool assay (OPA) markers allowed to assign the seven linkage groups to chromosomes $1S^{sh}$ to $7S^{sh}$. Comparison of the OPA marker sequences with the barley genome sequences indicated that chromosomes of both species are highly collinear [73], which is in agreement with normal homoeologous relationships of the S^{sh} genome chromosomes to wheat. On the other hand, a genetic map of 137 loci constructed in *Ae. speltoides* compared to that of *T. monococcum* confirmed that the S genome conserves the gross chromosome structure of wheat [74].

2.4. Other Aegilops Species

Exploration of the chromosome structure in other *Aegilops* species has been based on genetic mapping and, in some instance, on physical mapping of low copy DNA sequences. Construction of a linkage map in *Ae. umbellulata* (UU), which was based on RFLP loci previously mapped in bread wheat, revealed that multiple chromosome rearrangements occurred in the evolution of this species. The U genome underwent a minimum of 11 reciprocal translocations and inversions to arrive at its current structure [75]. Additional possible rearrangements involving chromosomes 2U, 5U and 6U were supported by macrosyntenic comparisons of wheat and *Ae. umbellulata* chromosomes based on the chromosomal assignment of Conserved Orthologous Set (COS) markers in both species [76].

The M genome of *Ae. comosa* underwent some rearrangements relative to wheat. Chromosome 2M carries some genetic material from 5M, and a segment from 7M is present on 3M [76]. However, the M genome poses a chromosome structure more similar to that of wheat than the U genome.

A FISH map of 121 cDNA probes, with an average of 18 markers per chromosome, revealed the presence of evolutionary rearrangements in *Ae. markgrafii* (syn. *Ae. caudata*, genome CC) [77]. Chromosomes 1C and 5C are collinear to wheat chromosomes 1D and 5D. The other five chromosomes are highly rearranged. The density of markers per chromosome arm made possible to determine the positions of at least 19 breakpoints, which flanked adjacent rearranged chromosome segments with preserved synteny. Rearrangements consisted of pericentric inversions (chromosomes 2C, 3C, 4C and 6C), intra-chromosomal translocations (2C, 3C, 6C and 7C) and inter-chromosomal translocations between 2C-4C, 3C-4C and 6C-7C. In addition, one duplicated segment was found on chromosomes 6C and 7C, and three deletions on chromosomes 4C and 6C.

2.5. Rye

Chromosome pairing analysis of *ph1b* wheat x rye hybrids carrying diagnostic arm-specific C-banding markers showed normal homoeology of the rye chromosome arms lRS, lRL, 2RL, 3RS, 4RS and 5RS to wheat chromosomes. In contrast, 2RS, 3RL, 4RL, 5RL, 6RS, 6RL, 7RS and 7RL paired with wheat chromosomes from a different homoeologous group, denoting they were involved in evolutionary translocations [78,79]. A segment of 2RS showed some homoeology to 6AS, 6BS and 6DS, a distal segment of 3RLwas translocated from 6RL, 4RL carries a distal segment from 6RS and an intercalary segment from 7RS, a long distal segment of 5RL was translocated from 4RL, 6RL carries genetic materials from 3RL and 7RL, chromosome arm 7RS carries a translocated segment from 5RL, and 7RL contains some genetic material from 2RS.

A genetic map of rye constructed with RFLP markers confirmed the implication of six chromosomes in evolutionary rearrangements. Only chromosome 1R preserves the ancestral structure. Six translocations and one inversion were proposed to occur [80]. Later, a high-density linear gene-order map was established from combining a high-throughput transcript map covering 72% of the rye genes with chromosome survey sequencing. Seventeen conserved chromosome segments collinear with barley and wheat genomes were defined, and the sequence of translocations needed to reach this structure was suggested [9].

2.6. Barley

The first analysis of the barley chromosome structure relative to wheat was based on the compared positions of over 100 RFLP loci in the linkage maps of *T. monococcum* and barley [81]. The synteny of the A^m and H genomes was shown to be highly conserved. They differ in the translocation T(4AL-5AL) present in *T. monococcum* and in two paracentric inversions that involve the long arm of group 1 and 4 chromosomes. A further reported genetic map constructed using 242 EST and 96 RFLP markers from barley confirmed structural differences detected between the long arms of group 4 and 5 chromosomes [82]. Assembly of 21,766 genes in a linear order and its comparison with high-density physical maps of the wheat A, B and D genomes, showed macro-synteny perturbation because of translocations involving chromosome arms 4AL, 5AL and 7BS, pericentric inversions produced in chromosomes 4A, 5A, 2B and 3B, and the absence of some homoeologous segments in chromosome arms 1AS, 1AL, 2AL, 2DL, 5AL, 5BL and 5DL [83]. In addition, a pericentric inversion was assumed to have occurred in chromosome 1H after divergence of the barley lineage. A high-quality reference genome assembly for barley was later reported [84].

2.7. Other Triticeae Species

Among wild *Triticeae* species, the crested wheatgrass, *Agropyron cristatum*, shows a high crossability with wheat and other species, and represents a potential source of genes controlling resistance to biotic and abiotic stresses. The homoeology relationships of the diploid *A. cristatum* (PP) chromosomes to wheat was analyzed by the FISH locations of 45 wheat full-length cDNA probes in the seven chromosomes of *A. cristatum* [85]. Chromosomes 1P, 3P and 5P seem to be collinear with wheat chromosomes, while the other four chromosomes show a collinearity distortion, such as a pericentric inversion on 4P, a paracentric inversion on 6PL, or a reciprocal translocation between 2PS and 4PL.

The genus *Leymus* includes about 30 allotetraploid species that share genomes Ns and Xm derived from *Psathyrostachys* and an unknown progenitor, respectively. Useful agronomically-important traits have been introduced in wheat, but the homoeologous relationships of the 14 *Leymus* chromosomes to wheat have not been identified. A genetic map containing 799 EST markers was constructed from fertile hybrids between both of the wild rye species *L. triticoides* and *L. cinereus*. Alignments of EST *Leymus* markers to barley EST maps evidenced the presence of a reciprocal translocation between 4NsL and 5NsL, which is absent in the Xsm genome [86].

Thinopyrum bessarabicum (JJ) is a diploid perennial maritime wheatgrass with genes for salt tolerance and diseases resistance that can be used in wheat breeding. The physical location of 1150 SNP markers into 36 segments of the seven J chromosomes allowed us to construct a high-density molecular markers map. Synteny analysis with wheat evidenced rearrangements in the J genome, such as a reciprocal translocation between 4JL and 5JL, an interchange between centromeric segments of 2J and 5J, an intra-chromosomal translocation on 6JL, and a paracentric inversion on 7JS [87].

Intermediate wheatgrass (*Thinopyrum intermedium*) is a highly productive cool-season forage grass with resistance to drought, frost and many pests and diseases of wheat and other cereals. In addition, this species was identified as a good candidate for domestication and breeding as a perennial grain. *Th. intermedium* is an allohexaploid (2n = 6x = 42) whose most likely progenitors are the diploid species of *Thinopyrum* and *Pseudoregneria*. There are several proposed genome designations, one of which was $J^{vs}J^{vs}J^rJ^r$StSt, where J^{vs} and J^r represent ancestral genomes of the extant *Th. bessarabicum* and *Th. elongatum* species, respectively, and St the genome of some diploid species of *Pseudoroegneria* from Eurasia [88]. Genotyping-by-sequencing was used to construct a genetic map containing 10,029 markers distributed among the 21 linkage groups of *Th. intermedium* [89]. Comparison of the 21 linkage groups of intermediate wheatgrass with the barley reference genome sequence evidenced three highly collinear homoeologous genomes syntenic with the barley genome with only one exception, a reciprocal translocation between two chromosomes homoeologous to 4H and 5H. This translocation was suggested to belong to the St genome, but the chromosome structure of the *Th. bessarabicum* genome is consistent with a translocation inherited from the donor of the J^{vs} genome.

2.8. Recurrence and Variable Frequency of Chromosome Rearrangements

Evolutionary chromosome rearrangements vary between the species studied. However, the reciprocal translocation between 4L-5L is present in different diploid and polyploid species. Because of the discovery of this translocation in wheat and rye, a possible monophyletic origin of this translocation in a common ancestor of both lineages was rejected, since the D genome of wheat, which is more closely related to the A genome than the R genome, does not carry this translocation [90]. This translocation is also absent in the genomes of the *Aegilops* species including *Ae. tauschii*, *Ae. speltoides* and *Ae. sharonensis*. This suggests independent origins for translocation 4L-5L in different lineages. Accordingly, although in rye and wheat the breakpoint positions are identical on 4RL and 4AL, they are different on 5RL and 5AL [91]. These sites were suggested to occupy chromosome rearrangements hotspots, as they coincide with inversion junctions produced in the *Triticeae* ancestor [91].

On the other hand, variation in the number of gross chromosome rearrangements among different species suggests that the chromosome structure does not evolve parallel to its genome differentiation. For example, despite the theory that *Hordeum* diverged much earlier than *Aegilops*, barley, *Ae. tauschii*, *Ae. speltoides* and *Ae. sharonensis* show a highly conserved synteny. In contrast, rye, *Ae. umbellulata* and *Ae. caudata*, all of which diverged later than *Hordeum*, underwent a considerable number of chromosome rearrangements. Two of these species, rye and *Ae. caudata*, have been shown to undergo introgressive hybridization during their evolution, which is considered the source of their highly rearranged genomes [9,77]. The role of interspecific hybridization on genome reorganization is also apparent in the formation of allopolyploids. Chromosome rearrangements produced in tetraploid wheats contrast the highly conserved synteny of their diploid progenitors. Allopolyploidy is usually accompanied by extensive genome reorganization and changes in gene expression within a short period of time in wheat and other plant lineages. Both the genetic control of pairing and the physical divergence of the homoeologous genomes are considered the genetic systems responsible for the cytological diploidization of polyploid wheat [92,93]. Consequently, interspecific hybridization appears as a releasing factor of genome reorganization, which may accelerate the diploidization of allopolyploids. When recurrent polyploid formation is accompanied with variable genome rearrangements, as in tetraploid wheats, structural chromosome differentiation may hinder the genetic flow between the polyploids of different origins and facilitate their speciation.

3. An Overview on Meiotic Recombination in Plants

Meiotic recombination is the cellular process that generates new allele combinations upon which natural or artificial selection can act to favor the establishment of better adapted genotypes, or more useful agronomical traits, respectively. Plant breeders search for novel varieties, which combine valuable traits present in different parental lines. The genetic information provided by both parents is reshuffled in the F1 hybrid meiosis, and genetic combinations of interest appear in the offspring. When traits of interest are controlled by non-linked genes, the searched gene combination is the result of random chromosome segregation at the anaphase I of the hybrid. In the case of joining alleles present in homologous (or homoeologous) chromosomes, intra-chromosomal homologous (or homoeologous) recombination is required. Intra-chromosomal homologous recombination is produced during the first meiotic division as a result of the repairing process triggered at leptotene by a programmed production of double-strand DNA breaks (DSBs) catalyzed by the topoisomerase-like SPO11 protein [94,95]. The repairing process is configured over an intact template that may reside in the sister chromatid or in a chromatid of the homologous partner. There is evidence that both pathways exist in meiosis, although meiotic chromosomes seem to be organized to facilitate the choice of homologous recombination [96]. Nucleo-filaments formed by 3′ single-strand DNA overhangs generated from each DSB bound to recombinases RAD51 and DMC1 to invade a double-strand DNA stretch of the homologous chromosome to find its complementary strand [97]. The use of a non-sister homologous template may culminate with either a reciprocal exchange of large homologous chromosome segments, i.e., a crossover (CO), or the non-reciprocal exchange of a small DNA sequence, i.e., a non-crossover (NCO).

The DSBs repairing process is broadly conserved among plants, and comprehensive insights on the underlying molecular mechanism have been provided in recent reviews [98–100]. Only a minor fraction (5% in *Arabidopsis*, or maize) of the homologous interactions are resolved as COs. This is enough to ensure the occurrence of at least one CO, the obligatory CO, per chromosome pair. Most COs are processed in the class I pathway, which relies on a group of proteins called ZMMs (SHOC1/ZIP2, HEI10, ZIP4, MER3, MSH4, MSH5 and PTD) and two additional proteins MLH1 and MLH3 not included in the ZMM group [98]. A feature of the class I COs is that they prevent the occurrence of additional COs nearby. Class II COs are dependent upon structure-specific endonucleases including MUS81, and are insensitive to interference [101]. In addition to these two pathways that lead to the CO production, other mechanisms are involved in DSBs repairing. Three other groups of proteins are involved in the restriction of the CO number [99,100,102]: (i) FANCM and its cofactors MHF1 and MHF2 are thought to unwind post-invasion intermediates to promote NCOs through the synthesis-dependent strand annealing (SDSA) pathway, (ii) the BLM/Sgs1 helicase homologs RECQ4A/RECQ4B and the associated proteins TOP3α and RMI1, which process probably different recombination intermediates of FANCM, and (iii) FIGL1 and its partner FLIP, which may control the activity of the recombinases RAD51 and DMC1. Mutations that disrupt any of these three NCOs promoter pathways increase the frequency of class II COs dependent of MUS81 in *Arabidopsis* [102]. Recombination frequency increases with the distance to the centromere, and reaches maximum values in distal regions.

Regular segregation of homologous chromosomes at anaphase I is facilitated by the formation of chiasmata, which are visible from diplotene to metaphase I and represent the cytological expression of COs produced in previous stages. The occurrence of a CO between homologous chromatids happens in a context in which each member of the partner and its sister chromatid are held together by a number of ring-shaped cohesin complexes scattered along the chromosomes axes and formed after DNA replication. The cohesin complexes also play an additional role in the DSBs repairing process [103]. Cohesion is released from chromosome arms at anaphase I, but not from the centromere region, where sister kinetochores remain associated and oriented to the same pole [104]. Connections between sister kinetochores are resolved at anaphase II, allowing their segregation to opposite poles.

In most species, COs are non-randomly distributed along chromosomes. They are clustered at recombination hotspots that alternate with poor recombination domains. Initial steps of recombination are to some extent responsible for the CO distribution, but it is not clear whether the CO landscape mirrors the non-uniform DSBs' positioning [105]. High-resolution maps of recombination events in plants show COs located in regions close to gene promoters and terminators [106]. While the DSB and CO maps are rather similar in *Arabidopsis* [107], they are very different in maize [108]. Only a quarter of DSBs produced in maize occur near genic regions, and can be processed as COs, the remaining DSBs situate in repetitive DNA, and do not form COs. Different patterns of CO distribution have been reported in plants. The most common situation is a pronounced localization in a distal euchromatic region. This is the case of maize [109] and *Triticeae* species [26,110–113], with an extremely distal location in *Ae. speltoides* [74]. Some species show a quite different pattern, for example, CO hotspots extend throughout the entire chromosome in *Arabidopsis* and rice [99], while most COs are located in the proximal quarter of the *Allium fistulosum* chromosomes [114].

Preferences for CO location in euchromatic regions suggest a specific role of chromatin structure in the recombination distribution. In fact, CO hotspots have low DNA methylation and transposons in some cereals [98]. In *Arabidopsis*, COs and recombination are correlated with active chromatin features, such as the modified histone H2A.Z, trimethylation of histone H3 on lysine 4 (H3K4me3), low DNA methylation, and low-nucleosome-density regions [115,116]. Epigenetic marks are also responsible for the absence of COs in centromeres and the surrounded heterochromatic regions of plant chromosomes.

Modification of pericentromeric epigenetic marks, such as histone 3 lysine 9 dimethylation (H3K9me2), and DNA methylation in CG and non-CG sequences in *Arabidopsis* defective mutants for the H3K9 methyltransferase genes KYP/SUVH4 SUVH5 SUVH6, or the CHG DNA methyltransferase gene CMT3, increases CO frequency in proximal regions, despite the fact that the total number of COs is not affected [117]. In addition to chromatin structure, the initiation of recombination changes across the telomere-centromere axis. DSBs appear initially in sub-telomeric CO-rich regions, and extend later to the intercalary CO-poor regions of barley chromosomes [113], which suggests that early recombination events are preferred for CO formation.

Homoeologous chromosomes are present in interspecific hybrids and allopolyploids, and represent potential partners to be involved in the DSBs' repairing process. Such chromosome partner selection was found to be suppressed in hexaploid wheat by the action of the *Ph1* locus of chromosome 5B [36,37]. *Ph1* was first assigned to a structure composed of a segment of heterochromatin inserted into a cluster of seven cyclin-like-dependent kinase (CDK) genes [118]. This *Ph1* assignment was later put into question by the meiotic phenotype of a different candidate gene, *C-Ph1*, which was silenced [119]. Included in the heterochromatin segment of chromosome 5B is the ZMM gene *ZYP4* (*TAZYP4-B2*). This gene was proposed to be *Ph1*, as supported by the high level of homoeologous pairing found in the hybrids of two *Tazyp4-B2* TILLING mutants and one CRISPR mutant of wheat with *Ae. variabilis* [120,121]. In addition, extensive analysis of RNA seq data in wheat, wheat x rye hybrids and triticale with and without *Ph1*, indicated that the silenced C-*ph1* gene corresponded to a copy present on chromosome 5D, since the copy on chromosome 5B is not expressed at any meiosis stage [122]. *Ph1* is the most effective suppressor gene of homoeologous recombination in polyploid wheats. *Ph2*, another suppressor located on chromosome arm 3DS, shows an intermediate effect [123]. Two main steps of bivalents' formation seem to be under the action of *Ph1*: The suppression of recombination between homoeologous chromosomes, despite the fact that they form synaptonemal complex (SC) during zygotene, and the correction of SC multivalents to form two or more bivalents during pachytene [124]. However, how this can be accomplished is poorly understood.

4. Modulating the Meiotic Recombination Landscape for Cereal Improvement

The reference sequence of the wheat genome gives access to 107,891 genes, with an uneven distribution along the chromosomes [26]. Intercalary CO-poor regions harbor a larger fraction of genes than the highly recombinogenic distal regions. A similar genome organization is present in barley [84]. Increasing recombination in chromosome regions with low CO frequency is upmost in importance from the breeding perspective. A possible strategy would be to increase the overall CO frequency using anti CO mutants. The TILLING mutants generated in tetraploid and hexaploid wheats, which are accessible in public databases [125], may be very useful for this purpose. Mutation of anti-CO genes greatly increase the CO frequency in *Arabidopsis* and in crops such as *Brassica*, rice, pea and tomato [106]. However, this strategy may fail in promoting CO in low-recombining regions of crops because, in *Arabidopsis*, extra COs fall into the highly recombining regions [102]. Approaches, such as manipulation of epigenetic marks associated with low recombining regions, or the induction of DSBs at specific chromosome sites by a fusion of SPO11 and different DNA-binding proteins, have been proposed to increase the CO number in intercalary or proximal regions [106]. Available stocks of wheat mutants [125] can also be used for the identification of candidate genes involved in chromatin organization, which might have some effect on meiotic recombination.

An alternative approach may be based on the environmental effect on meiotic development. A number of exogenous factors, including environmental stresses, agrochemical, heavy metals, combustible gases, pharmaceutical and pathogens, are known to affect meiosis in plants [126]. Many of these factors produce meiotic abnormalities such as laggards, univalents, bridges, stickiness, or precocious chromosome movements, which detract their practical value to modify the recombination pattern. Special attention has been received on the effect of temperature, which is species-specific. In *Arabidopsis*, the response to temperature variation follows a U-shaped curve with increased CO frequencies at low (8°) and high (28°) temperatures relative to a medium-range value (18°) [127]. Extra COs produced at extreme temperatures were of the class I pathway. In contrast, barley plants subjected at a temperature of 30° produce a lower number of chiasmata in male meiosis, which change their positions towards more proximal locations. Repositioning of recombination events affects only the class I COs [113,128]. Modification of nutrient concentration also affect the recombination frequency. Among *Triticeae*, a high phosphate level increases chiasma frequency in rye [129] and magnesium of the Hoagland's solution increases CO frequency in wheat and the hybrids of wheat lacking *Ph1* with related species [121]. However, it is unknown whether these nutrients modify the CO distribution or not.

Wild relatives are a source of variation for introgressing crops traits, providing tolerance to biotic and abiotic stresses. Introgression can be achieved through meiotic recombination between homologous chromosomes using wild species that share same genome with the crop. A number of genes providing resistance to diseases and pests were transferred from diploid and polyploid wild *Triticum* species to cultivated wheats [34,130]. Genetic variability is much higher in wild species containing genomes that are homoeologous to those of the crops. Transfer of useful genes from such species requires the induction of recombination between homoeologous chromosomes of the cultivated and wild species. This homoeologous recombination is induced in the absence of chromosome 5B, or using the deletion mutant *ph1b*, which lacks *Ph1* [131]. This mutant line has the disadvantage of accumulating translocations between homoeologous chromosomes produced by meiotic recombination [132]. However, other mutants of the *ZIP4-B2* gene in the *Ph1* locus recently obtained, form only homologous bivalents at metaphase I, and preserve better the genome stability [120]. Another way of inducing homoeologous recombination is through the use of genes that suppress the effect of *Ph1*. Suppressors of *Ph1* have been reported in *Ae. speltoides* and other species. One of the two major suppressor loci of *Ae. speltoides*, *Su1-Ph1*, which is located distally on the long arm of chromosome 3S, has been introgressed into chromosome 3A of hexaploid and tetraploid wheats [133] and can be used in interspecific gene transfer.

Pioneer works of wild introgression into wheat via homoeologous recombination were those transferring disease resistance genes from *Ae. comosa* and *Ag. elongatum* [134–136]. The transfer of resistance to the yellow rust *Puccinia striiformis* from *Ae. comosa* to wheat was induced by the *Ae. speltoides* genome. Plants homozygous for the resistant allele gave rice to the wheat variety known as Compare [134,135]. Recombination between wheat chromosomes 3D and 7D and their homoeologues of *Ag. elongatum*, produced in the absence of chromosome 5B, made possible the transfer to wheat of resistance to the leaf rust *Puccinia recondita* [136]. After these genetic transfers to wheat, many others have been produced. Species of the genus *Aegilops* represent a valuable source of genetic variability used in wheat improvement. Most *Aegilops* species were crossed with wheat to obtain amphiploids, and addition, substitution, translocation and segmental introgression lines. More than 40 resistance genes from the *Aegilops* species have been introgressed into wheat through chromosome translocation or homoeolgous recombination, and some of them have been used in wheat production [40]. An ample genetic variability present in perennial wild grasses and wild ryes has been incorporated also into the wheat genome in the form of amphiploids or derivatives such as addition, substitution or radiation translocation lines, as well as recombinant lines with segmental introgressions. Transfers involving *Thinopyrum* species have provided valuable contributions in wheat cultivar development [137].

Intercrops transfers have also been employed in some breeding programs. Many efforts have been made to use the gene pool of rye in wheat improvement. The production of hexaploid triticale AABBRR is a representative example of the success of combining the genomes of both species to produce an excellent feed crop. However, the highest potential of rye introgressions into wheat has been manifested in the production of a number of wheat cultivars carrying the wheat-rye translocations 1RS.1BL or 1RS.1AL, with a noteworthy positive effect on yield production [138]. Wheat-barley hybrids and introgressions have been produced also [139]. The transfer to wheat of barley genes controlling agronomical traits, such as drought tolerance, high β-glucan content, salt tolerance or earliness, is conditioned by the low crossability between both species. Efforts should be made to increase the efficiency of the crosses.

Hybrids between wheat and other *Hordeum* species have also been developed. *Tritordeum*, a *H. chilense* × durum wheat amphiploid, is the most successful introgression [140,141]. After the improvement of different agronomical traits in field experiments, *Tritordeum* became a synthetic cereal to some extent comparable to triticale. Recombination between the chromosomes of wheat and *H. chilense* has been induced in the absence of *Ph1*, supporting the idea that the transfer is possible using this approach [142].

In addition to the desirable gene, recombinant chromosomes carrying segments of wheat and alien chromosomes might carry other genes that reduce the agronomical value of the introgression. Therefore, primary recombinant chromosomes should be engineered to remove undesirable genes. This is possible after the production of sets of primary recombinant chromosomes formed by wheat-centromere-wheat/alien genetic material, which differ in the translocation breakpoint position, and their reciprocal counterparts, formed by alien-centromere-alien/wheat genetic material. Double heterozygotes containing chromosomes of the types wheat-centromere-wheat/alien and alien-centromere-alien/wheat, carrying the desirable gene from wild species, that undergo a CO in the overlapping alien region, produce secondary recombinants wheat-centromere-wheat/alien/wheat with the desirable gene into a shorter alien intercalary segment [143,144]. In the absence of such primary recombinant chromosome sets, the size of the translocated alien segment can be reduced in additional rounds of homoeologous recombination between the translocated alien segment and the standard wheat chromosome.

5. The Impact of Chromosome Rearrangements on Meiotic Recombination

CO frequency in *Triticeae* was estimated to range between two and three COs per chromosome, or more than one CO per chromosomal arm [112,145]. Distal confinement of the site of the first or only CO in each arm might be conditioned by the subtelomeric location of the initial interactions between homologs imposed by the bouquet arrangement [146]. In fact, distal recombinational events precede those more proximally located in barley [113]. CO interference could condition the formation and position of additional COs. Given the preferred distal localization of COs in many plant species, an alteration of the standard chromosome structure represents an approach used to understand the molecular basis underlying this recombination pattern. Chromosome structural mutants, such as deletions and inversions, which modify the relative position of the genetic material present in a given chromosome, have been produced in wheat, rye and *Arabidopsis*. In wheat, both types of chromosome mutants are viable in the homozygous and heterozygous conditions, and the same is with rye when the mutant chromosome is introgressed in wheat. Large deletions produced in *Arabidopsis* are viable only in heterozygotes [147]. Recombination in such heterozygotes denotes no major effect of the deletions in the total number of COs. The loss of COs at the deletion sites is compensated by increases in recombination frequencies elsewhere on the same chromosome. Thus, changes in the physical structure of a given chromosome redistribute the COs within that chromosome [147].

Studies using deletions and inversions in wheat and rye were aimed at verifying the ability of a given region to form CO after changing its position in the centromere-telomere axis. Heterozygotes for the loss of a long terminal segment of wheat chromosome arms 4AL, 2BL and 5BL underwent a considerable reduction of chiasma frequency at metaphase I in these arms, but the level of chiasmata became normal in homozygotes for the truncated arms [148]. On the other hand, homozygotes for the loss of the distal 25% of the 1BL arm, or the distal 41% of 5BL, increase the recombination rate of the middle arm region, without modifying that of the proximal region, relative to wild type plants [149,150]. These results suggest that the position of a segment in the telomere-centromere axis was a decisive factor in its capacity to produce a CO. However, other studies contradict this idea. Regardless, synapsis is completed, COs are infrequent in the proximal third of the rye chromosome arm 5RL, both in homozygotes for the standard chromosome 5R, and homozygotes lacking the distal 70% of its long arm [151]. Even more convincing is the effect of the repositioning from distal to proximal of the highly recombinogenic region of the arms 1RL of rye and 2BS and 4AL of wheat, as a result of large paracentric inversions. COs are restricted to the proximal region in all homomozygotes for the inversion [152–154].

Thus, CO distributions in the arms 4AL and 2BS of wheat and 1RL and 5RL of rye are not dependent of the distance to the telomere; other factors such the DNA sequence and the pattern of chromatin organization should condition the recombination landscape.

Given that the truncation of chromosome arms 1BL and 5BL increases the frequency of recombination in their middle region [149,150], a similar effect was expected in other chromosomes of wheat. Progressive shortening of the arms of wheat chromosomes 4A and those of the B genome by terminal deletions showed an apparent shift of the CO site from distal to middle-proximal regions [155]. Chiasmate associations at metaphase I of the truncated homologous arms estimated the CO frequency in the fraction of the arms present in homozygotes for terminal deletions of different size. According to the preferred distal localization of COs, a decrease in the CO frequency accompanied the progressive reduction of the arm length. The recombination level of the standard chromosome was not recuperated in most deletions, but a considerable increase in the middle chromosome arm regions was supported. Deletions with breakpoints in sub-distal sites in chromosome arms del2BS, del2BL and del6BS show higher CO frequencies than deleted arms with more distal breakpoints. Most of the COs produced in the intact chromosome 3B locate in the distal 68 Mb (~ 20% of the total arm) of 3BS and the distal 59 Mb (~ 14% of the total arm) of 3BL [112]. These chromosome segments are smaller than those missing in deletions 3BS-7 (25%) and 3BL-11 (19%). Lines for these two deletions show very high CO frequencies, 66% the 3BS-7 deletion line and 99% the 3BL-11 line [153], suggesting that the shortening of the chromosome arm increases the level of recombination in intercalary regions of chromosome 3B. The intercalary deletion of the *ph1b* mutant maps 0.9 cM from the centromere, which corresponds to a CO frequency of 0.018 [156]. Despite that the interval centromere-*ph1b* deletion is longer than the fraction of the 5BL arm present in the del5BL-5 truncated chromosome, homozygotes for the del5BL-5 deletion form at least one CO in the deleted arm in 62% of meiocytes. The increase of CO frequency in intercalary regions induced by terminal deletions seems to be common to most wheat chromosomes. The 1BS arm represents an exception, since different deletion lines show a low CO frequency. The reason of the CO redistribution in deletion lines is still unknown, but future research in this field can provide valuable information to modify the recombination landscape.

The frequency of association between the homoeologous arms of individual chromosomes of wheat, and between wheat and related species chromosomes, has been reported in interspecific hybrids of wheat with rye, *Ae. longissima*, *Ae. sharonensis*, *Ae. speltoides* or *T. timopheevii* [46,63,67–70,78,79]. The long arms of groups 1 and 2 chromosomes of wheat and rye differ in the amount of C-heterochromatin present in their subtelomeric regions. Such differences made possible to identify the parental heterochromatin constitution of chromosomes at anaphase I, as well as recombinant wheat-wheat (A-B and B-D) and wheat-rye (A-R, B-R and D-R) homoeologous chromosomes in *ph1b* ABDR hybrids. The recombinant chromosome ratio fits the frequency of association at metaphase I in all the different homoeologous arm combinations [78,157]. Thus the frequency of association at metaphase I between homoeologous arms represents a good estimate of the homoeologous recombination frequency.

Among bread wheat chromosomes, both arms of chromosome 4A, and the arms 5AL and 7BS are involved in evolutionary chromosome rearrangements. The same happens with rye chromosome arms 2RS, 3RL, 4RL, 5RL, 6RS, 6RL, 7RS and 7RL, which are involved in multiple translocations relative to the wheat D genome, as well as with the arms $4S^lL$ and $7S^lL$ of *Ae. longissima*. In both *ph1b* mutant wheat × rye and *ph1b* wheat × *Ae. longissima* hybrids, the frequencies of all possible associations wheat-wheat (A-B, A-D, and B-D), wheat-rye (W-R) and wheat-*Ae. longissima* (W-S^l), are known. Thus, it is possible to assess the effect of chromosome structure on homoeologous recombination by a comparison of the level of recombination between chromosome arms showing normal homoeologous relationships and those rearranged during evolution. Mean frequencies of wheat-wheat, wheat-rye and wheat-*Ae. longissima* associations involving arms with conserved macro-synteny, or arms with gross rearrangements, are given in Table 1.

Table 1. Average frequency (%) of association at metaphase I between syntenic and rearranged homoeologous arms in hybrids of the *ph1b* mutant wheat (W) with rye (R) or *Ae. longissima* (S^l).

Homoeologous Association	Syntenic Arms	Rearranged Arms	Reference
A-B-D	4.67	0.95	[44]
A-B	7.68	3.55	
A-D	60.52	12.05	
B-D	17.48	0	
W-R	10.76	3.51	[78]
W-S^l	56.86	25.3	[67]

Although there are differences between short and long arms, as well as between homoeologous groups, on average, chromosome rearrangements cause a considerable reduction of homoeologous recombination, and may represent an obstacle in the transfer of genes located in rearranged chromosome arms. Exceptions are the 5RL arm of rye [78] and the $4S^lL$ of *Ae. longissima* [67], which probably carry long translocated segments.

T. timopheevii chromosomes evolved a different structure from that of *T. durum* and *T. aestivum*, despite the theory that the donors of the A^t and A genomes, and of the G and B genomes, are considered to be *T. urartu* and *Ae. speltoides*, respectively, in both lineages [63,64]. All three species share translocation T(4AL-5AL). However, while the inversions in 4A and translocation T(4AL-7BS) occurred in *T. turgidum*, four different translocations involving the arms $3A^tL$, $4A^tL$, $6A^tS$, 1GS and 4GS appeared in *T. timopheevii*. The effect of structural differences on the frequency of the A^t-A and G-B associations in interspecific hybrids between these three species is shown in Table 2. There is a considerable reduction of the recombination frequency between arms showing differences in the chromosome structure. The effect is more apparent among the G-B associations, which, on average, are less frequent than the A^t-A associations.

Table 2. Average frequency (%) of association at metaphase I between syntenic and rearranged chromosome arms of the A and A^t genomes and the B and G genomes in *T. timopheevii* × *T. turgidum* (A^tAGB) and *T. timopheevii* × *T. aestivum* (A^tAGBD) hybrids.

Hybrids	Association Type	Syntenic Arms	Rearranged Arms	Reference
A^tAGB	A^t-A	93.19	54.87	[63]
	G-B	45.89	8.5	
A^tAGBD	A^t-A	89.16	57.7	
	G-B	27.93	7.9	

Superimposed to the effect of chromosome structural differentiation on homoeologous recombination is the effect of the level of genetic affinity between cultivated and wild species, which is a result of their phylogenetic relationships. Chromosomes of species very distant in their evolution have a lower frequency of recombination than chromosomes of species with a higher closeness. This is apparent when the results obtained in wheat × *Aegilops* and wheat × rye hybrids are compared (Table 1). Chromosomes of *Ae. longissima*, *Ae. sharonensis* and *Ae. speltoides*, which are more closely related to wheat than to rye chromosomes, recombine with wheat chromosomes, especially with those of the B genome, much more frequently than with rye chromosomes. Accordingly, A^t-A chiasmate associations are more frequent than B-G associations in AA^{t}BG and AA^{t}BGD wheat hybrids (Table 2). Both conservation of macro-synteny and genetic differentiation are factors conditioning interspecific introgressions.

6. Concluding Remarks

Modulating the recombination landscape is of supreme importance for crop breeding, since a considerable number of genes with agronomical relevance are located in crossover-poor chromosome regions. Truncation of wheat chromosomes has been shown as a very useful way of increasing the recombination frequency in their intercalary regions. This approach is, of course, inviable in diploid crops and, most likely, it does not have an immediate application in wheat production. However, it may represent a reference to investigate the reason why such regions produce more COs in the truncated chromosomes than in the standard ones. Homologous and homoeologous recombinations show similar distribution patterns, and therefore, achievements reached in the first are also of interest for the second. A structural differentiation of chromosomes from different species has a negative effect on meiotic recombination between such chromosomes. This is probably a consequence of disturbances on chromosome interactions caused by the absence of synteny. In such cases, introgressions should be carried out using other methods such as radiation. Thus, the chromosome structural organization of related genomes is of interest in designing strategies of the introgression of useful genes into crops. Recent advances in the knowledge of wheat and related species' genomes will facilitate a rapid progress of future research projects aimed to introgress useful agronomical traits into wheat and other crops. Genomic datasets for wheat and some related species, which are of public access (htps://plants.ensambl.org), make possible the analysis of any region with its gene content and the comparison among different species. Expression profiles are available in a considerable number of tissues, including meiotic cells. Thus, expression patterns of genes identified in different species can be compared to homoeologous genes of wheat. This reinforces polyploid wheat as a very powerful model system for study, propelling it into the mainstream of the global plant research community.

Author Contributions: T.N. designed and wrote the work.

Funding: This work was supported by grant AGL2015-67349-P from Dirección General de Investigación Científica y Técnica, Ministerio de Economía y Competitividad of Spain.

Conflicts of Interest: The author declares no conflict of interest.

References

1. Bernhardt, N. Taxonomic treatments of Triticeae and the wheat Genus *Triticum*. In *Alien Introgression in Wheat. Cytogenetics, Molecular Biology and Genomics*; Molnár-Láng, M., Ceoloni, C., Doležel, J., Eds.; Springer: Berlin/Heidelberg, Germany, 2015; pp. 1–19. [CrossRef]
2. Feldman, M.; Levy, A.A. Origin and evolution of wheat and related Triticeae species. In *Alien Introgression in Wheat. Cytogenetics, Molecular Biology, and Genomics*; Molnár-Láng, M., Ceoloni, C., Doležel, J., Eds.; Springer: Berlin/Heidelberg, Germany, 2015; pp. 21–76. [CrossRef]
3. Gaut, B.S. Evolutionary dynamics of grass genomes. *New Phytol.* **2002**, *154*, 15–28. [CrossRef]
4. Huang, S.; Sirikhachornkit, A.; Su, X.; Faris, J.; Gill, B.; Haselkorn, R.; Gornicki, P. Genes encoding plastid acetyl-CoA carboxylase and 3-phosphoglycerate kinase of the *Triticum/Aegilops* complex and the evolutionary history of polyploid wheat. *Proc. Natl. Acad. Sci. USA* **2002**, *99*, 8133–8138. [CrossRef]
5. Eilam, T.; Anikster, Y.; Millet, E.; Manisterski, J.; Feldman, M. Genome size in natural and synthetic autopolyploids and in a natural segmental allopolyploid of several Triticeae species. *Genome* **2009**, *52*, 275–285. [CrossRef] [PubMed]
6. Blattner, F.R. Progress in phylogenetic analysis and a new infrageneric classification of the barley genus *Hordeum* (Poaceae: Triticeae). *Breed. Sci.* **2009**, *59*, 471–480. [CrossRef]
7. The International Barley Genome Sequencing Consortium. A physical, genetic and functional sequence assembly of the barley genome. *Nature* **2012**, *491*, 711–716. [CrossRef] [PubMed]

8. Maraci, Ö.; Özkan, H.; Bilgin, R. Phylogeny and genetic structure in the genus *Secale*. *PLoS ONE* **2018**, *13*, e0200825. [CrossRef] [PubMed]
9. Martis, M.M.; Zhou, R.; Haseneyer, G.; Schmutzer, T.; Vrána, J.; Kubaláková, M.; Konig, S.; Kugler, K.G.; Scholz, U.; Hackauf, B.; et al. Reticulate evolution of the rye genome. *Plant Cell* **2013**, *25*, 3685–3698. [CrossRef] [PubMed]
10. Dvorak, J.; Akhunov, E.D. Tempos of gene locus deletions and duplications and their relationship to recombination rate during diploid and polyploid evolution in the *Aegilops-Triticum* alliance. *Genetics* **2005**, *171*, 323–332. [CrossRef] [PubMed]
11. Marcussen, T.; Sandve, S.R.; Heier, L.; Spannagl, M.; Pfeifer, M.; The International Wheat Genome Sequencing Consortium; Jakobsen, K.S.; Wulff, B.B.H.; Steuernagel, B.; Klaus, F.X.; et al. Ancient hybridizations among the ancestral genomes of bread wheat. *Science* **2014**, *345*. [CrossRef] [PubMed]
12. Gornicki, P.; Zhu, H.; Wang, J.; Challa, G.S.; Zhang, Z.; Gill, B.S.; Li, W. The chloroplast view of the evolution of polyploid wheat. *New Phytol.* **2014**, *204*, 704–714. [CrossRef] [PubMed]
13. Dvorak, J.; di Terlizzi, P.; Zhang, H.B.; Resta, P. The evolution of polyploid wheats: Identification of the A genome donor species. *Genome* **1993**, *36*, 21–31. [CrossRef] [PubMed]
14. Sarkar, P.; Stebbins, G.L. Morphological evidence concerning the origin of the B genome in wheat. *Am. J. Bot.* **1956**, *43*, 297–304. [CrossRef]
15. Dvorak, J.; Zhang, H.B. Variation in repeated nucleotide sequences sheds light on the phylogeny of the wheat B and G genomes. *Proc. Natl. Acad. Sci. USA* **1990**, *87*, 9640–9644. [CrossRef] [PubMed]
16. Kihara, H. Discovery of the DD-analyser, one of the ancestors of *Triticum vulgare*. *Agric. Hortic.* **1944**, *19*, 13–14.
17. McFadden, E.S.; Sears, E.R. The origin of *Triticum spelta* and its free-threshing hexaploid relatives. *J. Hered.* **1946**, *37*, 81–89. [CrossRef] [PubMed]
18. Ogihara, Y.; Tsunewaki, K. Diversity and evolution of chloroplast DNA in *Triticum* and *Aegilops* as revealed by restriction fragment analysis. *Theor. Appl. Genet.* **1988**, *76*, 321–332. [CrossRef] [PubMed]
19. Terachi, T.; Ogihara, Y.; Tsunewaki, K. The molecular basis of genetic diversity among cytoplasms of *Triticum* and *Aegilops*. 7. Restriction endonuclease analysis of mitochondrial DNA from polyploid wheats and their ancestral species. *Theor. Appl. Genet.* **1990**, *80*, 366–373. [CrossRef]
20. Miyashita, N.T.; Mori, N.; Tsunewaki, K. Molecular variation in chloroplast DNA regions in ancestral species of wheat. *Genetics* **1994**, *137*, 883–889.
21. Jiang, J.; Gill, B.S. New 18S-26S ribosomal RNA gene loci: Chromosomal landmarks for the evolution of polyploid wheats. *Chromosoma* **1994**, *103*, 179–185. [CrossRef]
22. Badaeva, E.D.; Friebe, B.; Gill, B.S. Genome differentiation in *Aegilops*. 1. Distribution of highly repetitive DNA sequences on chromosomes of diploid species. *Genome* **1996**, *39*, 293–306. [CrossRef]
23. Sasanuma, T.; Miyashita, N.T.; Tsunewaki, K. Wheat phylogeny determined by RFLP analysis of nuclear DNA. 3. Intra- and interspecific variations of five *Aegilops* Sitopsis species. *Theor. Appl. Genet.* **1996**, *92*, 928–934. [CrossRef] [PubMed]
24. Upadhya, M.D.; Swaminathan, M.S. Genome analysis in *Triticum zhukovskyi*, a new hexaploid wheat. *Chromosoma* **1963**, *14*, 589–600. [CrossRef]
25. International Wheat Genome Sequencing Consortium. A chromosome-based draft sequence of the hexaploid bread wheat *Triticum aestivum* genome. *Science* **2014**, *345*, 1251788. [CrossRef] [PubMed]
26. International Wheat Genome Sequencing Consortium. Shifting the limits in wheat research and breeding using a fully annotated reference genome. *Science* **2018**, *361*. [CrossRef]
27. Avni, R.; Nave, M.; Barad, O.; Baruch, K.; Twardziok, S.O.; Gundlach, H.; Hale, I.; Mascher, M.; Spannagl, M.; Wiebe, K.; et al. Wild emmer genome architecture and diversity elucidate wheat evolution and domestication. *Science* **2017**, *357*, 93–97. [CrossRef] [PubMed]
28. Ling, H.Q.; Ma, B.; Shi, X.; Liu, H.; Dong, L.; Sun, H.; Cao, Y.; Gao, Q.; Zheng, S.; Li, Y.; et al. Genome sequence of the progenitor of wheat A subgenome *Triticum urartu*. *Nature* **2018**, *557*, 424–428. [CrossRef] [PubMed]
29. Zhao, G.; Zou, C.; Li, K.; Wang, K.; Li, T.; Gao, L.; Zhang, X.; Wang, H.; Yang, Z.; Liu, X.; et al. The *Aegilops tauschii* genome reveals multiple impacts of transposons. *Nat. Plants* **2017**, *3*, 946–955. [CrossRef]
30. Kihara, H. Cytologische und genetische studien bei wichtigen getreidearten mit besonderer rücksicht auf das verhalten der chromosomen und die sterilität in den bastarden. *Mem. Coll. Sci. Kyoto Univ.* **1924**, *1*, 1–200.

31. Kihara, H. Genomanalyse bei *Triticum* und *Aegilops*. IX. Systematischer aufbau der gattung *Aegilops* auf genomanalytischer grundlage. *Cytologia* **1945**, *14*, 135–144. [CrossRef]
32. Lilienfeld, A.F.H. Kihara: Genome-analysis in *Triticum* and *Aegilops*. X. Concluding review. *Cytologia* **1951**, *16*, 101–123. [CrossRef]
33. Kihara, H.; Tanaka, M. Addendum to the classification of the genus *Aegilops* by means of genome-analysis. *Wheat Inform. Serv.* **1970**, *30*, 1–2.
34. Fedak, G. Alien Introgressions from wild Triticum species, T. monococcum, T. urartu, T. turgidum, T. dicoccum, T. dicoccoides, T. carthlicum, T. araraticum, T. timopheevii, and T. miguschovae. In *Allien Introgression in Wheat. Cytogenetics, Molecular Biology, and Genomics*; Molnár-Láng, M., Ceoloni, C., Doležel, J., Eds.; Springer: Berlin/Heidelberg, Germany, 2015; pp. 191–219.
35. Okamoto, M. Asynaptic effect of chromosome V. *Wheat Info. Serv.* **1957**, *5*, 6.
36. Sears, E.R.; Okamoto, M. Intergenomic chromosome relationship in hexaploid wheat. In Proceedings of the 10th International Congress of Genetics, Toronto, ON, Canada, 20–27 August 1958; pp. 258–259.
37. Riley, R.; Chapman, V. Genetic control of the cytologically diploid behaviour of hexaploid wheat. *Nature* **1958**, *182*, 713–715. [CrossRef]
38. Riley, R.; Chapman, V. The effect of the deficiency of the long arm of chromosome 5B on meiotic pairing in *Triticum aestivum*. *Wheat Info. Serv.* **1964**, *17*, 12–15.
39. Riley, R.; Kempana, C. The homoeologous nature of the non-homologous meiotic pairing in *Triticum aestivum* deficient for chromosome V (5B). *Heredity* **1963**, *18*, 287–306. [CrossRef]
40. Zhang, P.; Dundas, I.S.; McIntosh, R.A.; Xu, S.S.; Park, R.F.; Gill, B.S.; Friebe, B. Wheat–*Aegilops* Introgressions. In *Allien Introgression in Wheat. Cytogenetics, Molecular Biology, and Genomics*; Molnár-Láng, M., Ceoloni, C., Doležel, J., Eds.; Springer: Berlin/Heidelberg, Germany, 2015; pp. 221–243. [CrossRef]
41. Sears, E.R. *The Aneuploids of Common Wheat*; Research Bulletin No. 572; University of Missouri: Columbia, MO, USA, 1954; pp. 1–58.
42. Sears, E.R. Nullisomic-tetrasomic combinations in hexaploid wheat. In *Chromosome Manipulations and Plant Genetics*; Riley, R., Lewis, K.R., Eds.; Springer: Berlin/Heidelberg, Germany, 1966; Volume 20, pp. 29–45.
43. Okamoto, M. identification of the chromosomes of common wheat belonging to the A and B genomes. *Can. J. Genet. Cytol.* **1962**, *4*, 31–37. [CrossRef]
44. Naranjo, T.; Roca, A.; Goicoechea, P.G.; Giraldez, R. *Chromosome Structure of Common Wheat: Genome Reassignment of Chromosomes 4A and 4B*; Miller, T.E., Koebner, R.M.D., Eds.; Cambridge: Cambridge, UK, 1988; pp. 115–120.
45. Shepherd, K.W.; Islam, A.K.M.R. *Fourth Compendium of Wheat-Alien Chromosome lines*; Miller, T.E., Koebner, R.M.D., Eds.; Cambridge: Cambridge, UK, 1988; pp. 1373–1398.
46. Naranjo, T.; Roca, A.; Goicoechea, P.G.; Giraldez, R. Arm homoeology of wheat and rye chromosomes. *Genome* **1987**, *29*, 873–882. [CrossRef]
47. Danilova, T.V.; Friebe, B.; Gill, B.S. Development of a wheat single gene FISH map for analyzing homoeologous relationship and chromosomal rearrangements within the Triticeae. *Theor. Appl. Genet.* **2014**, *127*, 715–730. [CrossRef] [PubMed]
48. Anderson, J.A.; Ogihara, Y.; Sorrells, M.E.; Tanskley, S.D. Development of a chromosomal arm map for wheat based on RFLP markers. *Theor. Appl. Genet.* **1992**, *83*, 1035–1043. [CrossRef]
49. Devos, K.M.; Dubcovsky, J.; Dvořák, J.; Chinoi, C.N.; Gale, M.D. Structural evolution of wheat chromosomes 4A, 5A, and 7B and its impact on recombination *Theor. Appl. Genet.* **1995**, *91*, 282–288. [CrossRef]
50. Dvorak, J.; Wang, L.; Zhu, T.T.; Jorgensen, C.M.; Deal, K.R.; Dai, X.T.; Dawson, M.W.; Müller, H.-G.; Luo, M.-C.; Ramasamy, R.K.; et al. Structural variation and rates of genome evolution in the grass family seen through comparison of sequences of genomes greatly differing in size. *Plant J.* **2018**, *95*, 487–503. [CrossRef] [PubMed]
51. Naranjo, T. Chromosome structure of durum wheat. *Theor. Appl. Genet.* **1990**, *79*, 397–400. [CrossRef] [PubMed]
52. Dvorak, J.; Wang, L.; Zhu, T.; Jorgensen, C.M.; Luo, M.-C.; Deal, K.R.; Gu, Y.Q.; Gill, B.S.; Distelfeld, A.; Devos, K.M.; et al. Reassessment of the evolution of wheat chromosomes 4A, 5A, and 7B. *Theor. Appl. Genet.* **2018**, *131*, 2451–2462. [CrossRef] [PubMed]
53. Gill, B.S.; Friebe, B.; Endo, T.R. Standard karyotype and nomenclature system for description of chromosome bands and structural aberrations in wheat (*Triticum aestivum*). *Genome* **1991**, *34*, 830–883. [CrossRef]

54. Nelson, J.C.; Sorrells, M.E.; Van Deynze, A.E.; Lu, Y.H.; Atkinson, M.; Bernard, M.; Leroy, P.; Faris, J.D.; Anderson, J.A. Molecular mapping of wheat. Major genes and rearrangements in homoeologous groups 4, 5, and 7. *Genetics* **1995**, *141*, 721–731. [PubMed]
55. Endo, T.E.; Gill, B.S. The deletion stocks of common wheat. *J. Hered.* **1996**, *87*, 295–307. [CrossRef]
56. Mickelson-Young, L.; Endo, T.R.; Gill, B.S. A cytogenetic laddermap of the wheat homoeologous group-4 chromosomes. *Theor. Appl. Genet.* **1995**, *90*, 1007–1011. [CrossRef]
57. Miftahudin Ros, K.; Ma, X.-F.; Mahmoud, A.A.; Layton, J.; Rodriguez Milla, M.A.; Chikmawati, T.; Ramalingam, J.; Feril, O.; Pathan, M.S.; Surlan Momirovic, G.; et al. Analysis of expressed sequence tag loci on wheat chromosome group 4. *Genetics* **2004**, *168*, 651–663. [CrossRef]
58. Ma, J.; Stiller, J.; Wei, Y.; Zheng, Y.-L.; Devos, K.M.; Dolezel, J.; Liu, C. Extensive pericentric rearrangements in the bread wheat (*Triticum aestivum* L.) genotype 'Chinese Spring' revealed from chromosome shotgun data. *Genome Biol. Evol.* **2014**, *6*, 3039–3048. [CrossRef]
59. Jorgensen, C.; Luo, M.-C.; Ramasamy, R.; Dawson, M.; Gill, B.S.; Korol, A.B.; Distelfeld, A.; Dvorak, J. A high-density genetic map of wild emmer wheat from the Karaca dag region provides new evidence on the structure and evolution of wheat chromosomes. *Front. Plant. Sci.* **2017**, *8*, 1798. [CrossRef]
60. Salse, J.; Bolot, S.; Throude, M.; Jouffe, V.; Piegu, B.; Quraishi, U.M.; Calcagno, T.; Cooke, R.; Delseny, M.; Feuillet, C. Identification and characterization of shared duplications between rice and wheat provide new insight into grass genome evolution. *Plant Cell* **2008**, *20*, 11–24. [CrossRef] [PubMed]
61. Luo, M.C.; Deal, K.R.; Akhunov, E.D.; Akhunovaa, A.R.; Anderson, O.D.; Anderson, J.A.; Blake, N.; Clegg, M.T.; Coleman-Derr, D.; Conley, E.J.; et al. Genome comparisons reveal a dominant mechanism of chromosome number reduction in grasses and accelerated genome evolution in Triticeae. *Proc. Natl. Acad. Sci. USA* **2009**, *106*, 15780–15785. [CrossRef]
62. Jiang, J.; Gill, B.S. Different species-specific chromosome translocations in *Triticum timopheevii* and *T. turgidum* support the diphyletic origin of polyploid wheats. *Chromosome Res.* **1994**, *2*, 59–64. [CrossRef] [PubMed]
63. Maestra, B.; Naranjo, T. Structural chromosome differentiation between *Triticum timopheevii* and *T. turgidum* and *T. aestivum*. *Theor Appl. Genet.* **1999**, *98*, 744–750. [CrossRef]
64. Rodriguez, S.; Perera, E.; Maestra, B.; Diez, M.; Naranjo, T. Chromosome structure of *Triticum timopheevii* relative to *T. turgidum*. *Genome* **2000**, *43*, 923–930. [CrossRef] [PubMed]
65. Salina, E.A.; Leonova, I.N.; Efremova, T.T.; Röder, M.S. Wheat genome structure: Translocations during the course of polyploidization. *Funct. Integr. Genomics* **2006**, *6*, 71–80. [CrossRef] [PubMed]
66. Dobrovolskaya, O.; Boeuf, C.; Salse, J.; Pont, C.; Sourdille, P.; Bernard, M.; Salina, E. Microsatellite mapping of *Ae. speltoides* and map-based comparative analysis of the S, G, and B genomes of Triticeae species. *Theor. Appl. Genet.* **2011**, *123*, 1145–1157. [CrossRef]
67. Naranjo, T. Chromosome structure of *Triticum longissimum* relative to wheat. *Theor. Appl. Genet.* **1995**, *91*, 105–109. [CrossRef]
68. Naranjo, T.; Maestra, B. The effect of *ph* mutations on homoeologous pairing in hybrids of wheat with *Triticum longissimum*. *Theor. Appl. Genet.* **1995**, *91*, 1265–1270. [CrossRef]
69. Maestra, B.; Naranjo, T. Homoeologous relationships of Triticum sharonense chromosomes to T. aestivum. *Theor. Appl. Genet.* **1997**, *94*, 657–663. [CrossRef]
70. Maestra, B.; Naranjo, T. Homoeologous relationships of *Aegilops speltoides* chromosomes to bread wheat. *Theor. Appl. Genet.* **1998**, *97*, 181–186. [CrossRef]
71. Zhang, H.; Reader, S.M.; Liu, X.; Jia, J.Z.; Gale, M.D.; Devos, K.M. Comparative genetic analysis of the *Aegilops longissima* and *Ae. sharonensis* genomes with common wheat. *Theor. Appl. Genet.* **2001**, *103*, 518–525. [CrossRef]
72. Olivera, P.D.; Kilian, A.; Wenzl, P.; Steffenson, B.J. Development of a genetic linkage map for Sharon goatgrass (*Aegilops sharonensis*) and mapping of a leaf rust resistance gene. *Genome* **2013**, *56*, 367–376. [CrossRef] [PubMed]
73. Yu, G.; Champouret, N.; Steuernagel, B.; Olivera, P.D.; Simmon, J.; William, C.; Johnson, R.; Moscou, M.J.; Hernandez-Pinzon, I.; Green, P.; et al. Discovery and characterization of two new stem rust resistance genes in *Aegilops sharonensis*. *Theor. Appl. Genet.* **2017**, *130*, 1207–1222. [CrossRef] [PubMed]
74. Luo, M.-C.; Deal, K.R.; Yang, Z.-L.; Dvorak, J. Comparative genetic maps reveal extreme crossover localization in the *Aegilops speltoides* chromosomes. *Theor. Appl. Genet.* **2005**, *111*, 1098–1106. [CrossRef] [PubMed]

75. Zhang, H.; Jia, J.; Gale, M.D.; Devos, K.M. Relationships between the chromosomes of *Aegilops umbellulata* and wheat. *Theor. Appl. Genet.* **1998**, *96*, 69–75. [CrossRef]
76. Molnár, I.; Vrána, J.; Buresová, V.; Cápal, P.; Farkas, A.; Darkó, E.; Cseh, A.; Kubaláková, M.; Molnár-Láng, M.; Doležel, J. Dissecting the U, M, S and C genomes of wild relatives of bread wheat (*Aegilops* spp.) into chromosomes and exploring their synteny with wheat. *Plant J.* **2016**, *88*, 452–467. [CrossRef] [PubMed]
77. Danilova, T.V.; Akhunova, A.R.; Akhunov, E.D.; Friebe, B.; Gill, B.S. Major structural genomic alterations can be associated with hybrid speciation in *Aegilops markgrafii* (Triticeae). *Plant J.* **2017**, *92*, 317–330. [CrossRef]
78. Naranjo, T.; Fernández-Rueda, P. Homoeology of rye chromosome arms to wheat. *Theor. Appl. Genet.* **1991**, *82*, 577–586. [CrossRef]
79. Naranjo, T.; Fernández-Rueda, P. Pairing and recombination between individual chromosomes of wheat and rye in hybrids carrying the *ph1b* mutation. *Theor. Appl. Genet.* **1996**, *93*, 242–248. [CrossRef]
80. Devos, K.M.; Atkinson, M.D.; Chinoy, C.N.; Francis, H.A.; Harcourt, R.L.; Koebner, R.M.D.; Liu, C.J.; Masojć, P.; Xie, D.X.; Gale, M.D. Chromosomal rearrangements in the rye genome relative to that of wheat. *Theor. Appl. Genet.* **1993**, *85*, 673–680. [CrossRef] [PubMed]
81. Dubcovsky, J.; Luo, M.C.; Zhong, G.Y.; Bransteitter, R.; Desai, A.; Kilian, A.; Kleinhofs, A.; Dvorak, J. Genetic map of diploid wheat, *Triticum monococcum* L., and its comparison with maps of *Hordeum vulgare* L. *Genetics* **1996**, *143*, 983–999. [PubMed]
82. Hori, K.; Takehara, S.; Nankaku, N.; Sato, K.; Sasakuma, T.; Takeda, K. Barley EST markers enhance map saturation and QTL mapping in diploid wheat. *Breeding Sci.* **2007**, *57*, 39–45. [CrossRef]
83. Mayer, K.F.X.; Martis, M.; Hedley, P.E.; Šimková, H.; Liu, H.; Morris, J.A.; Steuernagel, B.; Taudien, S.; Roessner, S.; Gundlach, H.; et al. Unlocking the barley genome by chromosomal and comparative genomics. *Plant Cell* **2011**, *23*, 1249–1263. [CrossRef] [PubMed]
84. Mascher, M.; Gundlach, H.; Himmelbach, A.; Beier, S.; Twardziok, S.O.; Wicker, T.; Radchuk, V.; Dockter, C.; Hedley, P.E.; Russell, J.; et al. A chromosome conformation capture ordered sequence of the barley genome. *Nature* **2017**, *544*, 427–433. [CrossRef] [PubMed]
85. Said, M.; Hřibová, E.; Danilova, T.V.; Karafiátová, M.; Čížková, J.; Friebe, B.; Doležel, J.; Gill, B.S.; Vrána, J. The *Agropyron cristatum* karyotype, chromosome structure and cross-genome homoeology as revealed by fluorescence in situ hybridization with tandem repeats and wheat single-gene probes. *Theor. Appl. Genet.* **2018**, *131*, 2213–2227. [CrossRef] [PubMed]
86. Larson, S.R.; Kishii, M.; Tsujimoto, H.; Qi, L.; Chen, P.; Lazo, G.R.; Jensen, K.B.; Wang, R.R.C. *Leymus* EST linkage maps identify 4NsL–5NsL reciprocal translocation, wheat-*Leymus* chromosome introgressions, and functionally important gene loci. *Theor. Appl. Genet.* **2012**, *124*, 189–206. [CrossRef]
87. Grewal, S.; Yang, C.; Hubbart Edwards, S.; Scholefeld, D.; Ashling, S.; Burridge, A.J.; King, I.P.; King, J. Characterisation of *Thinopyrum bessarabicum* chromosomes through genome-wide introgressions into wheat. *Theor. Appl. Genet.* **2018**, *131*, 389–406. [CrossRef]
88. Wang, R.R.C.; Larson, S.R.; Jensen, K.B.; Bushman, B.S.; DeHaan, L.R.; Wang, S.; Yan, X. Genome evolution of intermediate wheatgrass as revealed by EST-SSR markers developed from its three progenitor diploid species. *Genome* **2015**, *58*, 63–70. [CrossRef]
89. Kantarski, T.; Larson, S.; Zhang, X.; DeHaan, L.; Borevitz, J.; Anderson, J.; Poland, J. Development of the first consensus genetic map of intermediate wheatgrass (*Thinopyrum intermedium*) using genotyping-by-sequencing. *Theor. Appl. Genet.* **2017**, *130*, 137–150. [CrossRef]
90. Naranjo, T. The use of homoeologous pairing in the identification of homoeologous relationships in Triticeae. *Hereditas* **1992**, *116*, 219–223. [CrossRef]
91. Li, W.; Challa, G.S.; Zhu, H.; Wei, W. Recurrence of chromosome rearrangements and reuse of DNA breakpoints in the evolution of the Triticeae genomes. *G3* **2016**, *6*, 3837–3847. [CrossRef] [PubMed]
92. Feldman, M.; Levy, A.A. Genome evolution due to allopolyploidization in wheat. *Genetics* **2012**, *192*, 763–774. [CrossRef] [PubMed]
93. Leitch, I.J.; Bennnett, M.D. Polyploidy in Angiosperms. *Trends Plant Sci.* **1997**, *2*, 470–476. [CrossRef]
94. Keeney, S.; Giroux, C.N.; Kleckner, N. Meiosis-specific DNA double- strand breaks are catalyzed by Spo11, a member of a widely conserved protein family. *Cell* **1997**, *88*, 375–384. [CrossRef]
95. Robert, T.; Vrielynck, N.; Mézard, C.; de Massy, B.; Grelon, M. A new light on the meiotic DSB catalytic complex. *Semin. Cell Dev. Biol.* **2016**, *54*, 165–176. [CrossRef] [PubMed]

96. Pradillo, M.; Santos, J.L. The template choice decision in meiosis: Is the sister important? *Chromosoma* **2011**, *120*, 447–454. [CrossRef] [PubMed]
97. Hunter, N.; Kleckner, N. The single-end invasion: An asymmetric intermediate at the double-strand break to double-holliday junction transition of meiotic recombination. *Cell* **2001**, *106*, 59–70. [CrossRef]
98. Mercier, R.; Mézard, C.; Jenczewski, E.; Macaisne, N.; Grelon, M. The molecular biology of meiosis in plants. *Annu. Rev. Plant Biol.* **2015**, *66*, 297–327. [CrossRef]
99. Lambing, C.; Franklin, F.C.H.; Wang, C.-J.R. Understanding and manipulating meiotic recombination in plants. *Plant Physiol.* **2017**, *173*, 1530–1542. [CrossRef]
100. Lambing, C.; Heckmann, S. Tackling plant meiosis: From model research to crop improvement. *Front. Plant Sci.* **2018**, *9*, 829. [CrossRef] [PubMed]
101. Kohl, K.P.; Sekelsky, J. Meiotic and mitotic recombination in meiosis. *Genetics* **2013**, *194*, 327–334. [CrossRef] [PubMed]
102. Fernandes, J.B.; Seguéla-Arnaud, M.; Larchevêque, C.; Lloyd, A.H.; Mercier, R. Unleashing meiotic crossovers in hybrid plants. *Proc. Natl. Acad. Sci. USA* **2018**, *15*, 2431–2436. [CrossRef] [PubMed]
103. Bolaños-Villegas, P.; De, K.; Pradillo, M.; Liu, D.; Makaroff, C.A. In favor of establishment: Regulation of chromatid cohesion in plants. *Front. Plant Sci.* **2017**, *8*, 846. [CrossRef] [PubMed]
104. Li, X.; Dawe, R.K. Fused sister kinetochores initiate the reductional division in meiosis I. *Nature Cell Biol.* **2009**, *11*, 1103–1108. [CrossRef]
105. Mézard, C.; Tagliaro Jahns, M.; Grelon, M. Where to cross? New insights into the location of meiotic crossovers. *Trends Genet.* **2015**, *31*, 393–401. [CrossRef] [PubMed]
106. Blary, A.; Jenczewski, E. Manipulation of crossover frequency and distribution for plant breeding. *Theor. Appl. Genet.* **2019**, *132*, 575–592. [CrossRef]
107. Choi, K.; Zhao, X.; Tock, A.J.; Lambing, C.; Underwood, C.J.; Hardcastle, T.J.; Serra, H.; Kim, J.; Cho, H.S.; Kim, J.; et al. Nucleosomes and DNA methylation shape meiotic DSB frequency in *Arabidopsis thaliana* transposons and gene regulatory regions. *Genome Res.* **2018**, *1*, 1–15. [CrossRef]
108. He, Y.; Wang, M.; Dukowic-Schulze, S.; Zhou, A.; Tiang, C.-L.; Shilo, S.; Sidhu, G.K.; Eichten, S.; Bradbury, P.; Springer, N.M.; et al. Genomic features shaping the landscape of meiotic double-strand-break hotspots in maize. *Proc. Natl. Acad. Sci. USA* **2017**, *114*, 12231–12236. [CrossRef]
109. Anderson, L.K.; Doyle, G.G.; Brigham, B.; Carter, J.; Hooker, K.D.; Lai, A.; Rice, M.; Stack, S.M. High-resolution crossover maps for each bivalent of *Zea mays* using recombination nodules. *Genetics* **2003**, *165*, 849–865.
110. Lukaszewski, A.J.; Curtis, C.A. Physical distribution of recombination in B-genome chromosomes of tetraploid wheat. *Theor. Appl. Genet.* **1993**, *86*, 121–127. [CrossRef] [PubMed]
111. Akhunov, E.D.; Goodyear, A.W.; Geng, S.; Qi, L.L.; Echalier, B.; Gill, B.S.; Miftahudin, J.; Gustafson, J.P.; Lazo, G.; Chao, S.; et al. The organization and rate of evolution of wheat genomes are correlated with recombination rates along chromosome arms. *Genome Res.* **2003**, *13*, 753–763. [CrossRef] [PubMed]
112. Choulet, F.; Alberti, A.; Theil, S.; Glover, N.; Barbe, V.; Daron, J.; Pingault, L.; Sourdille, P.; Couloux, A.; Paux, E.; et al. Structural and functional partitioning of bread wheat chromosome 3B. *Science* **2014**, *345*, 1249721. [CrossRef] [PubMed]
113. Higgins, J.D.; Perry, R.M.; Barakate, A.; Ramsay, L.; Waugh, R.; Halpin, C.; Armstrong, S.J.; Franklin, F.C.H. Spatiotemporal asymmetry of the meiotic program underlies the predominantly distal distribution of meiotic crossovers in barley. *Plant Cell* **2012**, *24*, 4096–4109. [CrossRef] [PubMed]
114. Albini, S.M.; Jones, G.H. Synaptonemal complex spreading in *Allium cepa* and *A. fistulosum*: I. The initiation and sequence of pairing. *Chromosoma* **1987**, *95*, 324–338. [CrossRef]
115. Choi, K.; Zhao, X.; Kelly, K.A.; Venn, O.; Higgins, J.D.; Yelina, N.E.; Hardcastle, T.J.; Ziolkowski, P.A.; Copenhaver, G.P.; Franklin, F.C.; et al. Arabidopsis meiotic crossover hot spots overlap with H2A.Z nucleosomes at gene promoters. *Nat. Genet.* **2013**, *45*, 1327–1336. [CrossRef] [PubMed]
116. Yelina, N.E.; Lambing, C.; Hardcastle, T.J.; Zhao, X.; Santos, B.; Henderson, I.R. DNA methylation epigenetically silences crossover hot spots and controls chromosomal domains of meiotic recombination in *Arabidopsis*. *Genes Dev.* **2015**, *29*, 2183–2202. [CrossRef] [PubMed]
117. Underwood, C.J.; Choi, K.; Lambing, C.; Zhao, X.; Serra, H.; Borges, F.; Simorowski, J.; Ernst, E.; Jacob, Y.; Henderson, I.R.; et al. Epigenetic activation of meiotic recombination near *Arabidopsis thaliana* centromeres via loss of H3K9me2 and non-CG DNA methylation. *Genome Res.* **2018**, *28*, 1–13. [CrossRef]

118. Griffiths, S.; Sharp, R.; Foot, T.N.; Bertin, I.; Wanous, M.; Reader, S.; Colas, I.; Moore, G. Molecular characterization of *Ph1* as a major chromosome pairing locus in polyploid wheat. *Nature* **2006**, *439*, 749–752. [CrossRef]
119. Bhullar, R.; Nagarajan, R.; Bennypaul, H.; Sidhu, G.K.; Sidhu, G.; Rustgi, S.; von Wettstein, D.; Gill, K.S. Silencing of a metaphase I specific gene present in the *Ph1* locus results in phenotype similar to that of the *Ph1* mutations. *Proc. Nat. Acad. Sci. USA* **2014**, *111*, 14187–14192. [CrossRef]
120. Rey, M.D.; Martín, A.M.; Higgins, J.; Swarbreck, D.; Uauy, C.; Shaw, P.; Moore, G. Exploiting the *ZIP4* homologue within the wheat *Ph1* locus has identified two lines exhibiting homoeologous crossover in wheat-wild relative hybrids. *Mol. Breed.* **2017**, *37*, 95. [CrossRef]
121. Rey, M.-D.; Martín, A.C.; Smedley, M.; Hayta, S.; Harwood, W.; Shaw, P.; Moore, G. Magnesium increases homoeologous crossover frequency during meiosis in *ZIP4* (*Ph1* gene) mutant wheat-wild relative hybrids. *Front. Plant Sci.* **2018**, *9*, 509. [CrossRef] [PubMed]
122. Martín, A.C.; Borrill, P.; Higgins, J.; Alabdullah, A.; Ramírez-González, R.H.; Swarbreck, D.; Uauy, C.; Shaw, P.; Moore, G. Genome-wide transcriprion during early wheat meiosis is indepencent of synapsis, ploidy level and the *Ph1* locus. *Front. Plant Sci.* **2018**, *9*, 1791. [CrossRef]
123. Mello-Sampayo, T. Genetic regulation of meiotic chromosome pairing by chromosome 3D of *Triticum aestivum*. *Nat. New Biol.* **1971**, *230*, 22–23. [CrossRef] [PubMed]
124. Naranjo, T.; Benavente, A. The mode and regulation of chromosome pairing in wheat-alien hybrids (*Ph* genes, an updated view). In *Allien Introgression in Wheat. Cytogenetics, Molecular Biology, and Genomics*; Molnár-Láng, M., Ceoloni, C., Doležel, J., Eds.; Springer: Berlin/Heidelberg, Germany, 2015; pp. 133–162. [CrossRef]
125. Krasilevaa, K.V.; Vasquez-Gross, H.A.; Howell, T.; Bailey, P.; Paraiso, F.; Clissold, L.; Simmonds, J.; Ramirez-Gonzalez, R.H.; Wang, X.; Borrill, P.; et al. Uncovering hiden variation in polyploid wheat. *Pro. Nat. Acad. Sci. USA* **2017**, *114*, 913–921. [CrossRef]
126. Fuchs, L.K.; Jenkins, G.; Phillips, D.W. Anthropogenic impacts on meiosis in plants. *Front. Plant Sci.* **2018**, *9*, 1429. [CrossRef] [PubMed]
127. Lloyd, A.; Morgan, C.; Franklin, C.; Bomblies, K. Plasticity of meiotic recombination rates in response to temperature in *Arabidopsis*. *Genetics* **2018**, *208*, 1409–1420. [CrossRef] [PubMed]
128. Phillips, D.; Jenkins, G.; Macaulay, M.; Nibau, C.; Wnetrzak, J.; Fallding, D.; Colas, I.; Oakey, H.; Waugh, R.; Ramsay, L. The effect of temperature on the male and female recombination landscape of barley. *New Phytol.* **2015**, *208*, 421–429. [CrossRef]
129. Bennett, M.D.; Rees, H. Induced variation in chiasma frequency in rye in response to phosphate treatments. *Genet. Res.* **1970**, *16*, 325–331. [CrossRef]
130. Börner, A.; Ogbonnaya, F.C.; Röder, M.S.; Rasheed, A.; Peryannan, S.; Lagudah, E.S. Aegilops tauschii introgresions in wheat. In *Alien Introgression in Wheat. Cytogenetics, Molecular Biology, and Genomics*; Molnár-Láng, M., Ceoloni, C., Doležel, J., Eds.; Springer: Berlin/Heidelberg, Germany, 2015; pp. 245–271. [CrossRef]
131. Sears, E.R. An induced mutant with homoeologous pairing in wheat. *Can. J. Genet. Cytol.* **1977**, *19*, 585–593. [CrossRef]
132. Sánchez-Morán, E.; Benavente, E.; Orellana, J. Analysis of karyotypic stability of homoeologous-pairing (ph) mutants in allopolyploid wheats. *Chromosoma* **2001**, *110*, 371–377. [CrossRef]
133. Li, H.; Deal, K.R.; Luo, M.-C.; Ji, W.; Distelfeld, A.; Dvorak, J. Introgression of the *Aegilops speltoides Su1-Ph1* suppressor into wheat. *Front. Plant Sci.* **2017**, *8*, 2163. [CrossRef]
134. Riley, R.; Chapman, V.; Johnson, R. Introduction of yellow rust resistance of *Aegilops comosa* into wheat by genetically induced homoeologous recombination. *Nature* **1968**, *217*, 383–384. [CrossRef]
135. Riley, R.; Chapman, V.; Johnson, R. The incorporation of alien disease resistance in wheat by genetic interference with the regulation of meiotic chromosome synapsis. *Genet. Res.* **1968**, *12*, 199–219. [CrossRef]
136. Sears, E.R. *Agropyron-Wheat Transfer Induced by Homoeologous Pairing*; Sears, E.R., Sears, L.M.S., Eds.; Agriculture Experiment Station: Arlington, VI, USA, 1973; pp. 191–199.
137. Ceoloni, C.; Kuzmanovic, L.; Forte, P.; Virili, M.E.; Bitti, A. Wheat-perennial Triticeae introgressions: Major achievements and prospects. In *Alien Introgression in Wheat. Cytogenetics, Molecular Biology, and Genomics*; Molnár-Láng, M., Ceoloni, C., Doležel, J., Eds.; Springer: Berlin/Heidelberg, Germany, 2015; pp. 273–313. [CrossRef]

138. Lukaszewski, A. Introgressions between wheat and rye. In *Alien Introgression in Wheat. Cytogenetics, Molecular Biology, and Genomics*; Molnár-Láng, M., Ceoloni, C., Doležel, J., Eds.; Springer: Berlin/Heidelberg, Germany, 2015; pp. 163–189. [CrossRef]
139. Molnár-Láng, M.; Linc, G. Wheat-barley hybrids and introgession lines. In *Alien Introgression in Wheat. Cytogenetics, Molecular Biology, and Genomics*; Molnár-Láng, M., Ceoloni, C., Doležel, J., Eds.; Springer: Berlin/Heidelberg, Germany, 2015; pp. 315–345. [CrossRef]
140. Martín, A.; Sánchez-Monge Laguna, E. A hybrid between *Hordeum chilense* and *Triticum turgidum*. *Cereal Res. Commun.* **1980**, *8*, 349–353.
141. Martín, A.; Sánchez-Monge Laguna, E. Cytology and morphology of the amphiploid *Hordeum chilense* x *Triticum turgidum* conv. Durum. *Euphytica* **1982**, *31*, 261–267. [CrossRef]
142. Rey, M.D.; Calderón, M.C.; Prieto, P. The use of the *ph1b* mutant to induce recombination between the chromosomes of wheat and barley. *Front. Plant Sci.* **2015**, *6*, 160. [CrossRef]
143. Sears, E.R. Transfer of alien genetic material to wheat. In *Wheat Science-Today and Tomorrow*; Evans, L., Peacock, W.J., Eds.; Cambridge University Press: Cambridge, UK, 1981; pp. 75–89.
144. Lukaszewski, A.J. Manipulation of homologous and homoeologous chromosome recombination in wheat. In *Plant Cytogenetics. Methods in Molecular Biology*; Kianian, S., Kianian, P., Eds.; Springer: Berlin/Heidelberg, Germany, 2016; pp. 77–89. [CrossRef]
145. Sallee, P.J.; Kimber, G. *An Analysis of the Pairing of Wheat Telocentric Chromosomes*; Ramanujan, S., Ed.; Springer: New Dehli, India, 1978; pp. 408–419.
146. Holm, P.B. Chromosome pairing and chiasma formation in allohexaploid wheat: *Triticum aestivum* analyzed by spreading of meiotic nuclei. *Carlsberg Res. Commun.* **1986**, *51*, 239–294. [CrossRef]
147. Ederveen, A.; Lai, Y.; van Driel, M.A.; Gerats, T.; Peters, J.L. Modulating crossover positioning by introducing large structural changes in chromosomes. *BMC Genomics* **2015**, *16*, 89. [CrossRef]
148. Curtis, C.A.; Lukaszewski, A.J.; Chrzastek, M. Metaphase-I pairing of deficient chromosomes and genetic mapping of deficiency breakpoints in wheat. *Genome* **1991**, *34*, 553–560. [CrossRef]
149. Jones, L.E.; Rybka, K.; Lukaszewski, A.J. The effect of a deficiency and a deletion on recombination in chromosome 1BL in wheat. *Theor. Appl. Genet.* **2002**, *104*, 1204–1208. [CrossRef]
150. Qi, L.L.; Friebe, B.; Gill, B.S. A strategy for enhancing recombination in proximal regions of chromosomes. *Chromosome Res.* **2002**, *10*, 645–654. [CrossRef]
151. Naranjo, T.; Valenzuela, N.T.; Perera, E. Chiasma frequency is region-specific and chromosome conformation-dependent in a rye chromosome added to wheat. *Cytogenet. Genome Res.* **2010**, *129*, 133–142. [CrossRef]
152. Lukaszewski, A.J. Unexpected behaviour of an inverted rye chromosome arm in wheat. *Chromosoma* **2008**, *117*, 569–578. [CrossRef]
153. Lukaszewski, A.J.; Kopecky, D.; Linc, G. Inversions of chromosome arms 4AL and 2BS in wheat invert the patterns of chiasma distribution. *Chromosoma* **2012**, *121*, 201–208. [CrossRef]
154. Valenzuela, N.T.; Perera, E.; Naranjo, T. Dynamics of rye chromosome 1R regions with high and low crossover frequency in homology search and synapsis development. *PLoS ONE* **2012**, *7*, e36385. [CrossRef]
155. Naranjo, T. Forcing the shift of the crossover site to proximal regions in wheat chromosomes. *Theor. Appl. Genet.* **2015**, *128*, 1855–1863. [CrossRef]
156. Sears, E.R. *Mutations in Wheat That Raise the Level of Meiotic Chromosome Pairing*; Gustafson, J.P., Ed.; Plenum Press: New York, NY, USA, 1984; pp. 295–300.
157. Naranjo, T.; Fernández-Rueda, P.; Goicoechea, P.G.; Roca, A.; Giráldez, R. Homoeologous pairing and recombination between the long arms of group 1 chromosomes in wheat x rye hybrids. *Genome* **1988**, *32*, 293–301. [CrossRef]

Review

Chromosome Engineering in Tropical Cash Crops

Pablo Bolaños-Villegas [1,2]

[1] Fabio Baudrit Agricultural Research Station, University of Costa Rica, La Garita, Alajuela 20101, Costa Rica; pablo.bolanosvillegas@ucr.ac.cr

[2] Jardín Botánico Lankester, Universidad de Costa Rica, Cartago P.O. Box 302-7050, Costa Rica

Received: 11 November 2019; Accepted: 16 December 2019; Published: 15 January 2020

Abstract: Tropical and subtropical crops such as coffee, cacao, and papaya are valuable commodities, and their consumption is a seemingly indispensable part of the daily lives of billions of people worldwide. Conventional breeding of these crops is long, and yields are threatened by global warming. Traditional chromosome engineering and new synthetic biology methods could be used to engineer new chromosomes, facilitate the transmission of wild traits to improve resistance to stress and disease in these crops, and hopefully boost yields. This review gives an overview of these approaches. The adoption of these approaches may contribute to the resilience of agricultural communities, lead to economic growth and secure the availability of key resources for generations to come.

Keywords: tropical cash crops; coffee; cacao; papaya; chromosome engineering; synthetic biology

---

1. Introduction

By the year 2050, the world population may reach 9 billion and the demand for food may grow by 70% [1]. Unfortunately, climate change may increase the frequency of drought and intense precipitation events as well as elevate temperatures, which may exacerbate food insecurity and instability caused by low agricultural diversity and the high intensity of agricultural inputs [1]. For instance, current estimates indicate that an increase of 1 °C might cause a 10% to 20% reduction in the world's production of maize [2]. In fact, meta-analyses of climate change and its impact suggest that by 2030, crop yields may be decreased 50% [3].

The livelihood of millions of smallholder farmers and the survival of several national economies in Africa, Latin America, and Asia depend on crops usually considered commodities, or minor or orphan crops; examples are coffee (*Coffea arabica* L.) and cacao (*Theobroma cacao* L.) [4–6]. These crops are also believed to be at risk due to climate change [4,6]. For a tropical crop such as cacao, which is mostly grown in West Africa, the maximum temperature tolerated (38 °C) could be exceeded during hot and dry *El Niño* years, and the dry season may be extended for one additional month [6]. In Colombia, rising temperatures, longer droughts, and excessive rainfall have reduced coffee yields by 30% since 2008 [7]. Also for coffee, rapid deforestation and climate change may lead to the extinction of many wild African species and the permanent loss of diversity for future plant breeding [4].

Papaya is considered an important tropical crop because of its high nutritional value [8], especially its high content of vitamin A, vitamin C, thiamine, folate, niacin, riboflavin, iron, potassium and calcium [8]. Papaya is also cultivated for papain, an important proteolytic enzyme present in the latex of fruits harvested before ripening [9]. This enzyme improves digestion and can be used to cure ulcers [9]. Papain is also used for softening wool, preparing protein for animal food, the manufacture of cosmetics (toilet soap, toothpaste, and shampoo), and brewing beer [9]. Unfortunately, damage to photosynthetic carbon assimilation due to environmental stress, including water deficits, can reduce the biomass production and net carbon assimilation in papaya [9]. Therefore, a better understanding of the physiology and reproductive biology of papaya may facilitate its adaptation to climate change [9].

Unlike in major crops, minor tropical crops do not benefit from large public and private breeding programs that stress genomic and marker-assisted recurrent selection but rather focus on simple methods that prioritize backcrossing [5]. Hence, new approaches are urgently needed [5]. The following article outlines potential new approaches.

2. Fertility and Enhanced Meiotic Pairing

The development of new plant varieties by conventional plant breeding is based on the selection of traits already present in plant species. It relies heavily on sexual reproduction to accumulate favorable alleles for tolerance and resistance to stress, nutritional quality, or other agronomic and horticultural traits. Alleles that contribute to tolerance to stress or other traits can be obtained from local germplasm, landraces, breeding lines, wild species, or related genera [10]. *Theobroma cacao* has an anchored genome sequence of 324.7 megabase pairs (Mb) for the B97-61/B2 Criollo genome, which comprises approximately 28,798 predicted genes [11] (Table 1). The actual number of protein-coding genes is close to 21,437 [11]. About 24% of the genome consists of transposable elements such as the Gaucho long-terminal repeat (LTR) retrotransposon [12]. In cacao, the improvement of and selection for desirable traits is believed to have also caused accelerated accumulation of deleterious mutations because of population bottlenecks that started 3600 years ago [12,13]. Some of the mutations are suspected to be due to the process of fertilization itself [13]. However, as compared with non-domesticated *Theobroma cacao* varieties such as *Marañón* (1.68×10^{-5}) and *Purús* (1.23×10^{-4}), in the varieties *Amelonado*, *Contamana*, *Criollo* and *Guianna*, the same domestication process may account for differentially high recombination rates (expressed in centiMorgans per megabase pairs [cM/Mb], 4.04×10^{-6} to 3.91×10^{-3}) [14]. Thus, selecting for high meiotic recombination rates during conventional breeding might help ameliorate the low individual fitness of modern varieties.

Little has been reported about the chromosomes and the genome of the 22 *Theobroma* species (Malvaceae) [15]. All *Theobroma* species have the same diploid chromosome number ($2n = 20$) and similar morphology and length (1–2 μm) [15]. Positive staining with DNA fluorescent stain chromomycin A3, which preferentially binds to G-C sequences, revealed the prophase chromosomes of *Theobroma cacao* and *T. grandiflorum* with two terminal heterochromatic bands that co-localize with a 45S rDNA site. Each chromosomal complement has a single 5S rDNA site in the proximal region of another chromosome pair [15]. Despite this apparent chromosome uniformity within the genus, meiotic analyses of a few cultivars of *T. cacao* have indicated the presence of univalents and multivalents, thus hinting at the existence of structural rearrangements in chromosomes [15].

An important consideration for breeding is that cacao has a long juvenile period, up to five years, so selection for fruit-related traits is expensive and time-consuming because the trees must be maintained for at least three more years in order to evaluate the pods visually [16]. Cacao tree plantations require a large area of land and labor force [16]. Because diseases are a persistent problem for cacao, improved disease resistance via breeding is imperative [17]. Each year, infection with a variety of pathogens severely cripples global cacao production, with about 30% of all pods destroyed before harvesting [17]. In West Africa, severe outbreaks of *Phytophthora* spp. can destroy all cacao pods on a farm [17]. The fungal pathogen *Moniliophthora perniciosa* causes Witches' broom disease, which seriously cripples cacao yields in Ecuador and Brazil [16]. Cacao is highly heterozygous and mostly involves outbreeding; thus, generating inbred lines from crosses is laborious, and doubled haploid lines are notoriously difficult to develop [16]. Nonetheless, there are a few self-compatible cacao clones, such as the Ecuadorean CCN51 [16] and Costa Rican CATIE-R1 (Figure 1A).

Figure 1. Tropical crops that may be amenable to genome editing and chromosome engineering, as actually grown in Costa Rica. (**A**) Cacao field plot of newly developed, self-compatible variety CATIE-R1 at CATIE, the Tropical Agricultural Research and Higher Education Center (Cartago province, Costa Rica). (**B**) Arabica coffee plants, Catuaí Yellow cultivar (León Cortés county, Costa Rica). (**C**) Papaya, Pococí hybrid (Alajuela province, Costa Rica). Credits, (**A–C**): Allan Mata-Quirós (CATIE, Costa Rica), Emmanuel López (Passiflora Coffee Farm, Costa Rica), Eric Mora–Newcomer (University of Costa Rica).

Cultivated coffee mostly corresponds to the species *Coffea canephora* Pierre ($2n$ = 22 chromosomes, 364 Mb) and *C. arabica* L. ($2n$ = 44 chromosomes), the latter likely due to hybridization between *C. canephora* and *C. eugenioides* [18]. The chromosome number for the genus *Coffea* is n = 11, which is common for most genera of the family Rubiaceae [19]. Most *Coffea* species are diploid ($2n$ = 22) and self-incompatible; however, *C. arabica* L. is the only polyploid species ($2n$ = 44) of the genus, and it is self-compatible [19]. The genome of *C. canephora* is reported to be 569.9 Mb [20] (Table 1) and consists of 25,574 protein-coding genes, of which 23 are unique caffeine *N*-methyltransferases [18]. Approximately 50% of the genome consists of transposable elements, especially LTR retrotransposons of the *Copia* group [18]. Although they are morphologically different, *Coffea* species share common meiotic traits, such as little variation in the number of chiasmata and a high frequency of bivalents, even in interspecific hybrids [19]. Genomic in situ hybridization with total genomic DNA from *Coffea arabica* and *C. canephora* showed high cross-hybridization to chromosome preparations from *C. arabica*, *C. canephora*, and a *C. liberica*-introgressed *C. arabica* genotype [21]. All chromosomes can be labeled with these probes, which may suggest a close genetic relation between *Coffea arabica* and *C. canephora* and between the *Ca* genome of *C. canephora* and the *Ea* subgenome (from *C. eugenioides*), which are present in *C. arabica* [21].

Approximately 60% of the world's total coffee output is from *Coffea arabica*. The species is cultivated in Latin America, Ethiopia, Kenya, and Tanzania (see Figure 1B), whereas *Coffea canephora* is cultivated in Southern Brazil, Central Africa and Southeast Asia [22]. One of the key limitations for the breeding of new coffee varieties is the poor genetic diversity of current cultivars of *Coffea arabica*. This bottleneck is likely due to the limited number of individuals from the cultivars Típica and Bourbon that were introduced in the American tropics to initiate commercial plantations at the beginning of the 17th century [22].

It typically takes 25 years to develop a new coffee variety [22]. Caffeine is an alkaloid synthesized by several plants, including coffee and cacao [18]. Caffeine is synthesized in the leaves of coffee, where it may function as an insecticide, and in fruits and seeds, where it may inhibit the germination of competitors [18]. In the Arabica variety Caturra, famous for its good beverage quality, the caffeine content is approximately 1.15% (which is high); unfortunately, Caturra is susceptible to infection with coffee leaf rust (*Hemileia vastatrix*) and coffee berry disease (*Colletotrichum kahawae*) [22].

Papaya has a small genome, 372 Mb (Table 1); it is diploid ($2n$ = 18) and has two sex chromosomes, X and Yh [23], which are believed to have evolved about 7 million years ago [24]. Papaya pachytene chromosomal domains, brightly stained with fluorescent DNA stain 4′,6-diamidino-2-phenylindole

(DAPI), account for approximately 17% of the total papaya genome, which suggests that it is largely euchromatic [25]. The genome is believed to contain 21,784 protein-coding genes [23].

Table 1. Resources and methods for the study and transformation of genomes in cacao, coffee, and papaya.

Species	Genome Database	Suggested Method for Engineering	Estimated Time for Engineering and Regeneration of Plants
T. cacao L., Criollo genotype (B97-61/B2)	The Cocoa Genome Hub, http://cocoa-genome-hub.southgreen.fr/ [11]	*Agrobacterium*-mediated transient transformation with CRISPR/Cas9 of cacao leaves and cotyledon cells [17].	Five years for seedlings to reach sexual maturity [16]. Time required to perform engineering is unknown.
C. canephora P., accession DH200-94	Coffee Genome Hub, http://coffee-genome.org [20]	Cocultivation of embryogenic calli of *Coffea canephora* with *Agrobacterium* [26].	Six to 8 years to reach sexual maturity in *Coffea arabica* [27]. In *C. canephora*, induction of primary calli takes 1 month and induction of embryogenic calli takes an additional eight months. The selection of transformants takes at least one month, whereas plantlet regeneration may take another month. Total minimum estimated time: 11 months [26].
C. papaya, 'Sun Up'	*Carica papaya* ASGPBv0.4, https://phytozome.jgi.doe.gov/pz/portal.html#!info?alias=Org_Cpapaya [28]	*Agrobacterium*-mediated transformation of somatic and zygotic embryos [29,30].	Four months for plantlets to reach maturity [8]. Transformation takes a long time [31]. Following field cultivation of line Tainung #2 for one year, papaya shoots were collected and rooted in vitro with indole-3-butyric acid (time was not reported) [31]. Shoots were then cultured for six weeks to harvest adventitious roots [31]. Roots were cultured in vitro for three months to obtain somatic embryos [31]. Embryos were cocultured for two days and selected for 80 days [31]. Regenerants were cultured for one week to induce shooting and transferred to vermiculite in the greenhouse [31]. Estimated time: approximately one year and seven months.

Papaya is a large herbaceous perennial plant with a short juvenile growth phase of only 4 months [8]. Once papaya reaches reproductive maturity, each leaf axil will develop flowers continuously [8]. It is a trioecious species, in that its flowers may develop into three sex configurations: female (X), male (Y or MSY) and hermaphrodite (Yh or HSY) [32]. Wild papaya features a 1:1 proportion of male to female plants (dioecious) and cultivated papaya a proportion of 2:3 hermaphrodite and 1:3 female plants (gynodioecious) [32]. Hermaphrodite flowers develop into elongated fruits with a small cavity (see Figure 1C), a trait that is sought-after commercially, as opposed to flowers from female plants, which are ovoid and have a large cavity [28,33]; thus, great effort has been placed on determining the sex of plants grown in the field. The X region is only 3.5 Mb, whereas the HSY and MSY regions are about 8.1 Mb and are located on chromosome 1 [32]. Notably, any combination of the HSY and MSY regions is inviable [32], which has been interpreted as meaning that the HSY and MSY regions may not carry the instructions required for embryo development [34]. HSY and MSY are extremely methylated and heterochromatic as compared with regions located on the X chromosome [34].

Transcriptome and whole-genome sequencing of papaya have revealed the existence of a putative gene called *SHORT VEGETATIVE PHASE (SVP)-LIKE*, which might be involved in male-hermaphrodite flower differentiation and is present in naturally all-hermaphrodite populations of papaya grown in Pingtung, Taiwan [33]. The *SVP-like* gene is expressed specifically in papaya plants that are male (carry the MSY chromosome) and hermaphrodite (carry the HSY chromosome) but not in female plants (carry the X chromosome) [33]. The MSY chromosome appears to code for an intact SVP-like protein with both MADS- and K-box domains, the HSY chromosome codes for a partial K-box domain, and the X chromosome may code for a comparatively shorter fragment [33]. Close analysis suggests that in the HSY chromosome, the *SVP-like* gene has two insertions of *copia*-like retrotransposons [33], and a single polymorphism allows for identifying each insertion [33]. Other genes located on the HSY chromosome are also believed to mediate the development of hermaphrodite flowers, such as *CHROMATIN ASSEMBLY FACTOR 1 SUBUNIT A-LIKE* (*CpCAF1AL*) and *SOMATIC EMBRYOGENESIS RECEPTOR KINASE* (*CpSERK*) [34].

3. Mechanisms of Meiosis and Fertility

The process of meiosis has two purposes: to generate the genetic diversity transmitted by the gametes and to ensure correct segregation of chromosomes [35]. Defects in recombination and meiosis cause sterility [35]. In *Arabidopsis*, meiotic recombination begins with the formation of double-strand breaks (DSBs) in DNA, which is initiated by a topoisomerase-like protein called SPO11-1 and associated proteins such as MTOPVIB, PRD1–3, and DFO [36]. These breaks are processed by several proteins including MRE11, RAD50, NBS1 (the MRX complex) and Exo1 for endonucleolytic cleavage and removal of SPO11, COM1 and CtIP for end processing and by RPA1A-D and RPA2-3 for binding of single-strand DNA breaks at 3′ ends and for recruiting recombinases [36]. Meiotic recombination results from the activity of RAD51, RAD51B-D, XRCC2, XRCC3, DMC1 and BRCA2 [36], among others.

Meiotic crossovers intermix homologous chromosomes to produce novel combinations of alleles that are transmitted to offspring [37]. The study of the formation of meiotic crossovers is important for activities related to plant breeding such as the detection and identification of quantitative trait loci and gene mapping [37]. Infertility results from defects in the formation of crossovers because the homologs segregate randomly during cell division [35]. Thus, at least one crossover must be formed per chromosome pair to ensure the formation of balanced and viable gametes [35].

Two major crossover formation pathways exist in plants. The major one depends on the ZMM protein complex (MSH4, MSH5, MER3, HEI10, ZIP4, SHOC1, PTD) plus MLH1/3 and produces interfering crossovers, whereas the minor pathway relies on MUS81 and produces non-interfering crossovers [36]. Crossovers are relatively rare events because active mechanisms limit their formation [37]. In the model plant *Arabidopsis*, three anti-crossover pathways rely on the activity of the proteins RECQ4, FANCM, and FIGL1 [37]. RECQ4 is the homologue of yeast Sgsp and is a DNA helicase; FANCM is also a DNA helicase (which requires cofactors MHF1-2), whereas FIGL1 encodes an AAA ATPase [37]. In *Arabidopsis*, changes in their activity cause notable changes; for instance, recombination was increased four-fold in the *recq4a*/*recq4b* mutant but increased three-fold in the *fancm*−/− mutant [37]. The *recq4a*/*recq4b*/*figl* mutant showed an impressive eight-fold increase in recombination [37]. Work in peas, tomato and rice has shown that the identification of allele mutants for meiotic anti-helicases may boost recombination in crops to facilitate introgression of agriculturally valuable alleles [37].

Plants have two and possibly three homologous recombination pathways: (1) the standard DSB repair (DSBR) pathway and (2) the synthesis-dependent strand annealing (SDSA) pathway [38]. DSBR and SDSA pathways share common steps in inducing breaks but differ in how the resulting displacement loop (D-loop) is resolved [38]. In the standard DSBR pathway, strand exchange results in the initial establishment of a double Holliday junction (dHJ), and subsequent resolution of the dHJ leads to crossover formation between homologous chromosomes [38]. In the SDSA pathway, the dHJ is dissolved, which results in the formation of a non-crossover (gene conversion event) [38]. In somatic plant cells, homologous recombination of DSBs is believed to proceed through the non-crossover SDSA pathway [38]. Plants have another HDR pathway, single-strand annealing (SSA), that repairs DSBs by using homologous sequences found within the same DNA sequence [38]. In the SSA pathway, an adjoining region of complementarity is used as the annealing site for the free ends of the break, and ends that are non-complementary are trimmed [38]. Thus, the SSA pathway leads to the loss of sequences located between the two repeat regions [38]. The FANCM helicase is believed to be involved in both the SDSA and SSA pathways [38], and its cofactors MHF1/2 are also considered to have a similar role in limiting crossovers and to act genetically in the same pathway [36] (see Figure 2A).

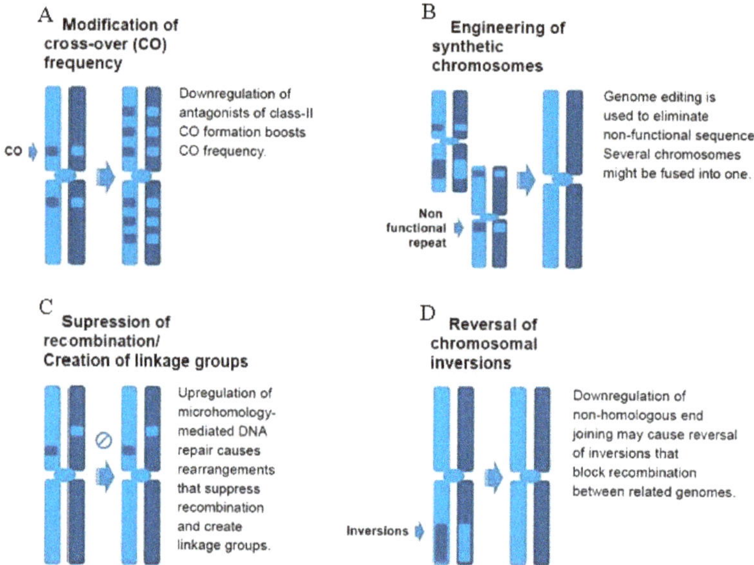

Figure 2. Possibilities for genome and chromosome engineering in tropical crops. (**A**) Modification of cross-over (CO) frequency, elimination or downregulation of meiotic antagonists of class II cross-over formation may lead to a significant increase in non-interfering cross-over formation, which may facilitate introgression of beneficial alleles from wild relatives. (**B**) Engineering of synthetic chromosomes, following the example of synthetic yeast strains, genome editing in plants may allow for the trimming of non-functional sequences, such as those corresponding to transposable elements, and facilitate the aggregation of open reading frames into so-called "Megachunks". Eventually, megachunks from different chromosomes may be combined into few, fully functional chromosomes. (**C**) Suppression of recombination and creation of linkage groups. In somatic tissues, downregulation of non-homologous end joining (NHEJ) by targeting either the Ku70/Ku80 heterodimer or the XRCC4 ligase leads to a shift to repair through backup EJ by DNA polymerase θ, which results in large deletions, inversions and translocations. These rearrangements may block recombination and lead to the de novo creation of new linkage groups that facilitate apomictic-like asexual reproduction. This trait may be useful for harvesting hybrid seeds that breed true to type in the next generation. (**D**) Downregulation of Ku70/Ku80 may lead to large chromosomal inversions up to 18 Kb in length, which may help undo chromosomal rearrangements that may occur during the evolution of new closely related species; such ancestral rearrangements may result in difficult recombination between chromosomes from commercial varieties and those from wild relatives. Downregulation of NHEJ may also facilitate CRISPR/Cas9-mediated genome editing.

Besides using homologous repair, plants may repair DSBs in their DNA by non-homologous end joining (NHEJ) via the Ku-dependent, "standard" NHEJ pathway or the highly erring Ku-independent backup-EJ pathway, especially in somatic cells, which are often the tissue of choice for genetic engineering experiments [38]. The backup-EJ pathway is also called microhomology-mediated end joining [38]. In canonical NHEJ, the Ku70-Ku80 heterodimer and the XRCC4 ligase recognize DSBs and bind to them, with minimal end processing, which results in minimal DNA loss (1 to 4 nt) [38]. In backup EJ, DSBs are bound by poly (ADP-ribose) polymerase proteins and the MRX complex in order to promote end-resection and generate homology between the two DNA sequences; however, the broken DNA ends are resected and extended by the very error-prone DNA polymerase θ, which results in large deletions, inversions, and translocations [38]. Downregulation of NHEJ genes is believed to shift most DNA repair toward homologous recombination and thus increase the efficiency of CRISPR/Cas9 editing [39,40].

Other possibilities for chromosome engineering have been suggested for *S. cerevisae* [41] and could be considered for engineering synthetic chromosomes in plants. For instance, all sub-telomeric repeats and retrotransposons could be removed until chromosomes can be organized into regular megachunks composed of fully functional open reading frames [41]. Methods could be developed to control the chromatin state of entire chromosomes [42] and even to fuse all chromosomes into only one as done in yeast with CRISPR/Cas9 [43] (see Figure 2B).

4. Possibilities for Genome Editing in Coffee, Cacao and Papaya

Genome editing is a technological breakthrough that allows for the creation of targeted mutations and the replacement of sequences with high specificity [26]. The CRISPR/Cas9 editing technology is inspired by bacterial adaptive immune systems that rely on clustered regularly interspaced, short palindromic repeats (CRISPR)-associated protein 9 (Cas9) whose function is to protect against invading foreign DNA by cleaving it with an RNA-guide [26]. The system is simple and efficient [26] and requires a trans-activating crRNA sequence (tracrRNA) to assist in the maturation of CRISPR RNAs (crRNAs) [44].

The most basic gene editing process by CRISPR relies on the formation of DSBs and appears to have 2 stages [45]. The first stage is recognition of a palindrome (PAM) [45,46]. This stage is determined by the sequence of PAM itself, by interactions with Cas9 interactions and by chromatin accessibility [45]. The second stage is the formation of the R-loop, which is a structure between the target DNA (taDNA) and the signal guide RNA (sgRNA) [45]. Usually, the formation of an RNA-DNA hybrid helix 20 bp long is required for sgRNA-target recognition and Cas9 cleavage, but shorter sequences (protospacers) may also work. The Cas9 endonuclease may tolerate mismatches in the distal region of the PAM (non-seed region), although perfect base-pairing in the proximal region of the PAM-proximal region (seed region) is preferred [45]. The PAM motif that Cas9 may recognize is 5′-NGG-3′ (where N = A, T, C, or G), but unfortunately, this G-rich sequence makes it difficult to design sgRNAs located within T-rich regions [44]. An alternative is to resort to Cpf1, a class 2 endonuclease in the CRISPR system, that is efficient in plant genome editing. Quite conveniently, the Cpf1 endonuclease does not require a tracrRNA sequence to form a mature crRNA, and it can also recognize T-rich PAM sequences. Additionally, cutting by the Cpf1 endonuclease produces cohesive ends, unlike Cas9, which produces blunt ends [44]. In addition to generating insertions and deletions at target sequences, CRISPR/Cas systems have been developed for base editing [44]. These base editors consist of a sgRNA-guided Cas9 nickase (e.g., nCas9) fused with a deaminase that causes very precise C-T or A-G base conversions [44].

In contrast to animal cells, plant zygotic cells are difficult to transform with CRISPR/Cas9 directly due to technological limitations [46]. Currently, the transformation of most plant species involves callus cells obtained by tissue culture and exposure of those cells to *Agrobacterium* [46]. The efficiency of transformation may depend on the choice of codons for Cas9 translation and the promoters used for the expression of Cas9 and the sgRNAs [46]. An exciting recent development is prime editing, a "search-and-replace" technology that facilitates targeted insertions, deletions, and base-to-base conversions without the need for DSBs [47]. This is possible because of prime editors (heavily modified Cas9 proteins) that feature a reverse transcriptase (RT) fused to an RNA-programmable nickase and a prime-editing guide RNA (pegRNA), all working together to copy genetic information from the pegRNA into the target sequence [47].

For cacao, cells isolated from somatic embryo cotyledons and young leaves (Scavina 6 variety) can be transiently transformed by using *Agrobacterium* carrying a CRISPR/Cas9 construct [17] (Table 1). Apparently, stable expression was achieved with the vector pGSh16.1010, which carries the Cauliflower Mosaic Virus (CaMV) *35S* promoter [17]. Coffee has been successfully transformed by using embryogenic callus cultures as the tissue of choice for transformation with *Agrobacterium tumefaciens* (strain LBA1119) [48] (Table 1). For the *Coffea canephora* clone 197, the *C. canephora U6* promoter was used to transform 6-month-old embryogenic calli derived from leaf segments for 10 min with a solution containing the *Agrobacterium tumefaciens* strain EHA105, selected for tolerance to cefotaxime and hygromycin and grown into plants [26]. A pCAMBIA 5300 binary vector carrying the Cas9 sequence was successfully used for these transformants [26]. Approximately 7% of all transformants tested homozygous for mutations of the phytoene desaturase gene (*CcPDS, Cc04_ g00540*) [26].

For papaya, genetic engineering has mostly involved biolistic-mediated transformation related to tolerance to the papaya ringspot virus, a Potyvirus that is quite possibly the most serious viral disease of papaya worldwide [29,30]. Nonetheless, *Agrobacterium*-mediated transformation of somatic and zygotic embryos has also been reported [30] (Table 1). CRISPR/Cas9-mediated genome editing against several viruses simultaneously in papaya may be technically feasible and within reach [30].

5. Chromosome Structure Editing in Crops

Besides the targeted editing of key genes, genome engineering may also allow for the modification of chromosome structure [40]. Theoretically, if several DSBs in DNA are introduced into the genome, chromosomes might be modified in a directed manner to create new combinations of fragments [40]. For instance, if two DSBs are induced on the same chromosome, deletions or inversions in the area in between may give rise to intrachromosomal rearrangements [40]. However, to create interchromosomal rearrangements, the formation of two or more DSBs on different chromosomes would be required [40]. Depending on the type of tissue, a somatic or meiotic crossover may be formed by inducing breaks on one or both homologues [40].

In *Arabidopsis*, inversions of about 18 kb in length were successfully passed on to offspring by using an egg cell-specific promoter (*EC1.1*) for Cas9 (SaCas9) expression [49]. Inversion and deletion frequencies were reported to be greater in a *ku70-1* mutant than the wild type, which suggests increased efficiency by mutagenic microhomology-mediated backup EJ [49]. This process has been observed in mice as well [50]. In agriculture, this approach may help reverse natural inversions between related species to facilitate outcrossing and transmission of beneficial alleles [49,50] (see Figure 2C). For example, in papaya, crosses of species from the related genus *Vasconcella* might promote transmission of tolerance to ringspot virus and the fungus *Phytophthora palmivora*, which causes fruit rot and destroys the root meristems [51]. However, meiosis is usually defective in these wild species [51], and lagging chromosomes and polyads are observed [51].

Another yet theoretical possibility for crops would be to induce an inversion to create new linkage groups and purposely suppress recombination [40] (see Figure 2D). Suppression of meiotic recombination is also observed during apomixis, a form of reproduction characterized by deregulation of meiosis during embryo sac formation [35], development of the embryo in a fertilization-independent way [35], and formation of viable endosperm in a fertilization-dependent or -independent way [35]. Apomixis is an effective strategy to retain genome-wide parental heterozygosity; recently, apomixis was successfully engineered in rice to facilitate clonal propagation of hybrids by seed [52].

6. Concluding Remarks

Advances in understanding agricultural genomes in tropical crops such as coffee, cacao, and papaya may hopefully allow for adaptation to climate change and increased output. It is important to take timely advantage of the possibilities brought about by genome editing technologies in order to boost recombination and crossover formation and as a result, improve the transmission of novel traits. It may also be possible for the first time to take advantage of the downregulation of NHEJ to create novel linkage groups and even to reverse natural chromosomal inversions between related species. Farmers in developing countries and their customers in developed economies may all benefit immensely from the adoption of these technologies.

Funding: This research was funded by Vicerrectoría de Investigación (University of Costa Rica) grant numbers B8069, B6602 and B5A52. And the APC was funded directly by Vicerrectoría de Investigación.

Acknowledgments: Work at the Bolaños-Villegas laboratory was supported by grants B8069, B6602, B5A52 and B5A49 from Vicerrectoría de Investigación (University of Costa Rica), which also covered publication costs for this article. Pablo is a young member affiliate of TWAS/UNESCO and a member of the American Society of Plant Biologists. This manuscript was kindly edited by Ms. Laura Smales (BioMedEditing, Toronto, Canada). Images were kindly edited by E. Bolaños-Villegas.

Conflicts of Interest: The author declares no conflicts of interest. The sponsors had no role in the design, execution, interpretation, or writing of the study.

References

1. Dhankher, O.P.; Foyer, C.H. Climate Resilient Crops for Improving Global Food Security and Safety. *Plant Cell Environ.* **2018**, *41*, 877–884. [CrossRef]
2. Tito, R.; Vasconcelos, H.L.; Feeley, K.J. Global Climate Change Increases Risk of Crop Yield Losses and Food Insecurity in the Tropical Andes. *Glob. Chang. Biol.* **2018**, *24*, e592–e602. [CrossRef] [PubMed]
3. Campbell, B.M.; Hansen, J.; Rioux, J.; Stirling, C.M.; Twomlow, S. Wollenberg. Urgent Action to Combat Climate Change and Its Impacts (SDG 13): Transforming Agriculture and Food Systems. *Curr. Opin. Environ. Sustain.* **2018**, *34*, 13–20. [CrossRef]
4. Davis, A.P.; Chadburn, H.; Moat, J.; O'Sullivan, R.; Hargreaves, S.; Lughadha, E.N. High Extinction Risk for Wild Coffee Species and Implications for Coffee Sector Sustainability. *Sci. Adv.* **2019**, *5*, eaav3473. [CrossRef] [PubMed]
5. Farrell, A.D.; Rhiney, K.; Eitzinger, A.; Umaharan, P. Climate Adaptation in a Minor Crop Species: Is the Cocoa Breeding Network Prepared for Climate Change? *Agroecol. Sustain. Food Syst.* **2018**, *42*, 812–833. [CrossRef]
6. Schroth, G.; Läderach, P.; Martinez-Valle, A.I.; Bunn, C.; Jassogne, L. Vulnerability to Climate Change of Cocoa in West Africa: Patterns, Opportunities and Limits to Adaptation. *Sci. Total Environ.* **2016**, *556*, 231–241. [CrossRef] [PubMed]
7. Van der Vossen, H.; Bertrand, B.; Charrier, A. Next Generation Variety Development for Sustainable Production of Arabica Coffee (*Coffea Arabica*, L.): A Review. *Euphytica* **2015**, *204*, 243–256. [CrossRef]
8. Liao, Z.; Yu, Q.; Ming, R. Development of Male-Specific Markers and Identification of Sex Reversal Mutants in Papaya. *Euphytica* **2017**, *213*, 53. [CrossRef]
9. Campostrini, E.; Schaffer, B.; Ramalho, J.D.C.; González, J.C.; Rodrigues, W.P.; Da Silva, J.R.; Lima, R.S.N. Environmental Factors Controlling Carbon Assimilation, Growth, and Yield of Papaya (*Carica Papaya* L.) Under Water-Scarcity Scenarios. In *Water Scarcity and Sustainable Agriculture in Semiarid Environment*; Elsevier: Amsterdam, The Netherlands, 2018; pp. 481–500. [CrossRef]
10. Gaur, R.K.; Verma, R.K.; Khurana, S.M.P. Genetic Engineering of Horticultural Crops. In *Genetic Engineering of Horticultural Crops*; Academic Press: Cambridge, MA, USA, 2018; pp. 23–46. [CrossRef]
11. Argout, X.; Martin, G.; Droc, G.; Fouet, O.; Labadie, K.; Rivals, E.; Aury, J.M.; Lanaud, C. The Cacao Criollo Genome v2.0: An Improved Version of the Genome for Genetic and Functional Genomic Studies. *BMC Genom.* **2017**, *18*, 730. [CrossRef]
12. Argout, X.; Salse, J.; Aury, J.M.; Guiltinan, M.J.; Droc, G.; Gouzy, J.; Allegre, M.; Chaparro, C.; Legavre, T.; Maximova, S.N.; et al. The Genome of Theobroma Cacao. *Nat. Genet.* **2011**, *43*, 101–108. [CrossRef]
13. Cornejo, O.E.; Yee, M.-C.; Dominguez, V.; Andrews, M.; Sockell, A.; Strandberg, E.; Livingstone, D.; Stack, C.; Romero, A.; Umaharan, P.; et al. Population Genomic Analyses of the Chocolate Tree, *Theobroma cacao*, L., Provide Insights into Its Domestication Process. *Commun. Biol.* **2018**, *1*, 1–16. [CrossRef] [PubMed]
14. Schwarzkopf, E.J.; Motamayor, J.C.; Cornejo, O.E. Genetic Differentiation and Intrinsic Genomic Features Explain Variation in Recombination Hotspots among Cocoa Tree Populations. *bioRxiv* **2018**. [CrossRef]
15. Dantas, L.G.; Guerra, M. Chromatin Differentiation between Theobroma Cacao L. and T. Grandiflorum Schum. *Genet. Mol. Biol.* **2010**, *33*, 94–98. [CrossRef] [PubMed]

16. Wickramasuriya, A.M.; Dunwell, J.M. Cacao Biotechnology: Current Status and Future Prospects. *Plant Biotechnol. J.* **2018**, *16*, 4–17. [CrossRef]
17. Fister, A.S.; Landherr, L.; Maximova, S.N.; Guiltinan, M.J. Transient Expression of CRISPR/Cas9 Machinery Targeting TcNPR3 Enhances Defense Response in Theobroma Cacao. *Front. Plant Sci.* **2018**, *9*, 268. [CrossRef]
18. Denoeud, F.; Carretero-Paulet, L.; Dereeper, A.; Droc, G.; Guyot, R.; Pietrella, M.; Zheng, C.; Alberti, A.; Anthony, F.; Aprea, G.; et al. The Coffee Genome Provides Insight into the Convergent Evolution of Caffeine Biosynthesis. *Science* **2014**, *345*, 1181–1184. [CrossRef]
19. Pinto-Maglio, C.A.F. Cytogenetics of Coffee. *Braz. J. Plant Physiol.* **2006**, *18*, 37–44. [CrossRef]
20. Dereeper, A.; Bocs, S.; Rouard, M.; Guignon, V.; Ravel, S.; Tranchant-Dubreuil, C.; Poncet, V.; Garsmeur, O.; Lashermes, P.; Droc, G. The Coffee Genome Hub: A Resource for Coffee Genomes. *Nucleic Acids Res.* **2015**, *43*, D1028–D1035. [CrossRef]
21. Herrera, J.C.; D'Hont, A.; Lashermes, P. Use of Fluorescence in Situ Hybridization as a Tool for Introgression Analysis and Chromosome Identification in Coffee (*Coffea Arabica* L.). *Genome* **2007**, *50*, 619–626. [CrossRef]
22. Moncada, M.D.P.; Tovar, E.; Montoya, J.C.; González, A.; Spindel, J.; McCouch, S. A Genetic Linkage Map of Coffee (*Coffea Arabica* L.) and QTL for Yield, Plant Height, and Bean Size. *Tree Genet. Genomes* **2016**, *12*, 1–17. [CrossRef]
23. Ming, R.; Hou, S.; Feng, Y.; Yu, Q.; Dionne-Laporte, A.; Saw, J.H.; Senin, P.; Wang, W.; Ly, B.V.; Lewis, K.L.T.; et al. The Draft Genome of the Transgenic Tropical Fruit Tree Papaya (*Carica Papaya* Linnaeus). *Nature* **2008**, *452*, 991–997. [CrossRef] [PubMed]
24. VanBuren, R.; Ming, R. Dynamic Transposable Element Accumulation in the Nascent Sex Chromosomes of Papaya. *Mob. Genet. Elem.* **2013**, *3*, e23462. [CrossRef] [PubMed]
25. Zhang, W.; Jiang, J. Molecular Cytogenetics of Papaya. In *Genetics and Genomics of Papaya, Plant Genetics and Genomics: Crops and Models 10*; Ming, R., Moore, P.H., Eds.; Springer Science + Business Media: New York, NY, USA, 2014; pp. 157–168. [CrossRef]
26. Breitler, J.C.; Dechamp, E.; Campa, C.; Zebral Rodrigues, L.A.; Guyot, R.; Marraccini, P.; Etienne, H. CRISPR/Cas9-Mediated Efficient Targeted Mutagenesis Has the Potential to Accelerate the Domestication of Coffea Canephora. *Plant Cell Tissue Organ Cult.* **2018**, *134*, 383–394. [CrossRef]
27. Sera, T. Coffee Genetic Breeding at IAPAR. *Crop Breed. Appl. Biotechnol.* **2001**, *1*, 179–199. [CrossRef]
28. Jiménez, V.M.; Mora-Newcomer, E.; Gutiérrez-Soto, M.V. Biology of the Papaya Plant. In *Genetics and Genomics of Papaya, Plant Genetics and Genomics: Crops and Models 10*; Ming, R., Moore, P.H., Eds.; Springer Science + Business Media: New York, NY, USA, 2014; pp. 17–33. [CrossRef]
29. Geetika, S.; Ruqia, M.; Harpreet, K.; Neha, D.; Shruti, K.; Singh, S.P. Genetic Engineering in Papaya. In *Genetic Engineering of Horticultural Crops*; Academic Press: Cambridge, MA, USA, 2018; pp. 137–154. [CrossRef]
30. Hamim, I.; Borth, W.B.; Marquez, J.; Green, J.C.; Melzer, M.J.; Hu, J.S. Transgene-Mediated Resistance to Papaya Ringspot Virus: Challenges and Solutions. *Phytoparasitica* **2018**, *46*, 1–18. [CrossRef]
31. Kung, Y.J.; Yu, T.A.; Huang, C.H.; Wang, H.C.; Wang, S.L.; Yeh, S.D. Generation of Hermaphrodite Transgenic Papaya Lines with Virus Resistance via Transformation of Somatic Embryos Derived from Adventitious Roots of in Vitro Shoots. *Transgenic Res.* **2010**, *19*, 621–635. [CrossRef]
32. VanBuren, R.; Zeng, F.; Chen, C.; Zhang, J.; Wai, C.M.; Han, J.; Aryal, R.; Gschwend, A.R.; Wang, J.; Na, J.K.; et al. Origin and Domestication of Papaya Yh Chromosome. *Genome Res.* **2015**, *25*, 524–533. [CrossRef]
33. Chen, J.R.; Urasaki, N.; Matsumura, H.; Chen, I.C.; Lee, M.J.; Chang, H.J.; Chung, W.C.; Ku, H.M. Dissecting the All-Hermaphrodite Phenomenon of a Rare X Chromosome Mutant in Papaya (*Carica Papaya* L.). *Mol. Breed.* **2019**, *39*, 14. [CrossRef]
34. Lee, C.Y.; Lin, H.J.; Viswanath, K.K.; Lin, C.P.; Chang, B.C.H.; Chiu, P.H.; Chiu, C.T.; Wang, R.H.; Chin, S.W.; Chen, F.C. The Development of Functional Mapping by Three Sex-Related Loci on the Third Whorl of Different Sex Types of *Carica papaya* L. *PLoS ONE* **2018**, *13*, e0194605. [CrossRef]
35. Fayos, I.; Mieulet, D.; Petit, J.; Meunier, A.C.; Périn, C.; Nicolas, A.; Guiderdoni, E. Engineering Meiotic Recombination Pathways in Rice. *Plant Biotechnol. J.* **2019**, 1–16. [CrossRef]
36. Wang, Y.; Copenhaver, G.P. Meiotic Recombination: Mixing It up in Plants. *Annu. Rev. Plant Biol.* **2018**, *69*, 577–609. [CrossRef] [PubMed]
37. Mieulet, D.; Aubert, G.; Bres, C.; Klein, A.; Droc, G.; Vieille, E.; Rond-Coissieux, C.; Sanchez, M.; Dalmais, M.; Mauxion, J.-P.; et al. Unleashing Meiotic Crossovers in Crops. *Nat. Plants* **2018**, *4*, 1010–1016. [CrossRef] [PubMed]

38. Que, Q.; Chen, Z.; Kelliher, T.; Skibbe, D.; Dong, S.; Chilton, M.-D. Nd Their Applications in Genome Engineering. In *Plant Genome Editing with CRISPR Systems*; Qi, Y., Ed.; Springer Science + Business Media, LLC: New York, NY, USA, 2019; pp. 3–24. [CrossRef]
39. Que, Q.; Chilton, M.D.M.; Elumalai, S.; Zhong, H.; Dong, S.; Shi, L. Repurposing Macromolecule Delivery Tools for Plant Genetic Modification in the Era of Precision Genome Engineering. In *Methods in Molecular Biology*; Springer Science + Business Media, LLC: New York, NY, USA, 2019; pp. 3–18. [CrossRef]
40. Schmidt, C.; Schindele, P.; Puchta, H. From Gene Editing to Genome Engineering: Restructuring Plant Chromosomes via CRISPR/Cas. *aBIOTECH* **2019**. [CrossRef]
41. Richardson, S.M.; Mitchell, L.A.; Stracquadanio, G.; Yang, K.; Dymond, J.S.; DiCarlo, J.E.; Lee, D.; Huang, C.L.V.; Chandrasegaran, S.; Cai, Y.; et al. Design of a Synthetic Yeast Genome. *Science* **2017**, *355*, 1040–1044. [CrossRef]
42. Ostrov, N.; Beal, J.; Ellis, T.; Gordon, D.B.; Karas, B.J.; Lee, H.H.; Lenaghan, S.C.; Schloss, J.A.; Stracquadanio, G.; Trefzer, A.; et al. Technological Challenges and Milestones for Writing Genomes. *Science* **2019**, *366*, 310–312. [CrossRef]
43. Shao, Y.; Lu, N.; Xue, X.; Qin, Z. Creating Functional Chromosome Fusions in Yeast with CRISPR–Cas9. *Nat. Protoc.* **2019**, *14*, 2521–2545. [CrossRef]
44. Xu, J.; Hua, K.; Lang, Z. Genome Editing for Horticultural Crop Improvement. *Hortic. Res.* **2019**, *6*, 113. [CrossRef]
45. Xu, X.; Duan, D.; Chen, S.J. CRISPR-Cas9 Cleavage Efficiency Correlates Strongly with TargetsgRNA Folding Stability: From Physical Mechanism to off-Target Assessment. *Sci. Rep.* **2017**, *7*, 143. [CrossRef]
46. Feng, C.; Su, H.; Bai, H.; Wang, R.; Liu, Y.; Guo, X.; Liu, C.; Zhang, J.; Yuan, J.; Birchler, J.A.; et al. High-Efficiency Genome Editing Using a Dmc1 Promoter-Controlled CRISPR/Cas9 System in Maize. *Plant Biotechnol. J.* **2018**, *16*, 1848–1857. [CrossRef]
47. Anzalone, A.V.; Randolph, P.B.; Davis, J.R.; Sousa, A.A.; Koblan, L.W.; Levy, J.M.; Chen, P.J.; Wilson, C.; Newby, G.A.; Raguram, A.; et al. Search-and-Replace Genome Editing without Double-Strand Breaks or Donor DNA. *Nature* **2019**, *576*, 149–157. [CrossRef]
48. Ribas, A.F.; Dechamp, E.; Champion, A.; Bertrand, B.; Combes, M.C.; Verdeil, J.L.; Lapeyre, F.; Lashermes, P.; Etienne, H. Agrobacterium-Mediated Genetic Transformation of *Coffea arabica* (L.) Is Greatly Enhanced by Using Established Embryogenic Callus Cultures. *BMC Plant Biol.* **2011**, *11*, 92. [CrossRef] [PubMed]
49. Schmidt, C.; Pacher, M.; Puchta, H. Efficient Induction of Heritable Inversions in Plant Genomes Using the CRISPR/Cas System. *Plant J.* **2019**, *98*, 577–589. [CrossRef] [PubMed]
50. Park, C.Y.; Sung, J.J.; Kim, D.W. Genome Editing of Structural Variations: Modeling and Gene Correction. *Trends Biotechnol.* **2016**, *34*, 548–561. [CrossRef] [PubMed]
51. Narducci Da Silva, E.; Neto, M.F.; Pereira, T.N.S.; Pereira, M.G. Meiotic Behavior of Wild Caricaceae Species Potentially Suitable for Papaya Improvement. *Crop Breed. Appl. Biotechnol.* **2012**, *12*, 52–59. [CrossRef]
52. Khanday, I.; Skinner, D.; Yang, B.; Mercier, R.; Sundaresan, V. A Male-Expressed Rice Embryogenic Trigger Redirected for Asexual Propagation through Seeds. *Nature* **2019**, *565*, 91. [CrossRef]

Article

The Confirmation of a Ploidy Periclinal Chimera of the Meiwa Kumquat (*Fortunella crassifolia* Swingle) Induced by Colchicine Treatment to Nucellar Embryos and Its Morphological Characteristics

Tsunaki Nukaya [1,†], Miki Sudo [1,†], Masaki Yahata [1,*], Tomohiro Ohta [1], Akiyoshi Tominaga [1], Hiroo Mukai [1], Kiichi Yasuda [2] and Hisato Kunitake [3]

1 Faculty of Agriculture, Shizuoka University, Shizuoka 422-8529, Japan; t_nukaya54@yahoo.co.jp (T.N.); sudo.miki@shizuoka.ac.jp (M.S.); tomohiro1_ota@pref.shizuoka.lg.jp (T.O.); tominaga.akiyoshi@shizuoka.ac.jp (A.T.); mukai.hiroo@shizuoka.ac.jp (H.M.)
2 School of Agriculture, Tokai University, Kumamoto 862-8652, Japan; yk964422@tsc.u-tokai.ac.jp
3 Faculty of Agriculture, University of Miyazaki, Miyazaki 889-2192, Japan; hkuni@cc.miyazaki-u.ac.jp
* Correspondence: yahata.masaki@shizuoka.ac.jp; Tel.: +81-54-641-9500
† These authors contributed equally to the article.

Received: 26 August 2019; Accepted: 12 September 2019; Published: 18 September 2019

Abstract: A ploidy chimera of the Meiwa kumquat (*Fortunella crassifolia* Swingle), which had been induced by treating the nucellar embryos with colchicine, and had diploid (2n = 2x = 18) and tetraploid (2n = 4x = 36) cells, was examined for its ploidy level, morphological characteristics, and sizes of its cells in its leaves, flowers, and fruits to reveal the ploidy level of each histogenic layer. Furthermore, the chimera was crossed with the diploid kumquat to evaluate the ploidy level of its reproductive organs. The morphological characteristics and the sizes of the cells in the leaves, flowers, and fruits of the chimera were similar to those of the tetraploid Meiwa kumquat and the ploidy periclinal chimera known as "Yubeni," with diploids in the histogenic layer I (L1) and tetraploids in the histogenic layer II (L2) and III (L3). However, the epidermis derived from the L1 of the chimera showed the same result as the diploid Meiwa kumquat in all organs and cells. The sexual organs derived from the L2 of the chimera were significantly larger than those of the diploid. Moreover, the ploidy level of the seedlings obtained from the chimera was mostly tetraploid. In the midrib derived from the L3, the chimera displayed the fluorescence intensity of a tetraploid by flow cytometric analysis and had the same size of the cells as the tetraploid and the Yubeni. According to these results, the chimera is thought to be a ploidy periclinal chimera with diploid cells in the outermost layer (L1) and tetraploid cells in the inner layers (L2 and L3) of the shoot apical meristem. The chimera had desirable fruit traits for a kumquat such as a thick pericarp, a high sugar content, and a small number of developed seeds. Furthermore, triploid progenies were obtained from reciprocal crosses between the chimera and diploid kumquat.

Keywords: anatomy; citrus; flow cytometry; histogenic layer; polyploidy breeding

1. Introduction

The shoot apical meristem of higher plants consists of three histogenic layers [1]. This is known as the "Tunica-Corpus" theory; i.e., the shoot apical meristem consists of L1 (Tunica) and L2/L3 (Corpus). In the genus *Citrus* and its related genera, the histogenic layers are differentiated by their parts: the dermal system (guard cell and juice sac) in L1, parenchyma and reproductive organs (mesophyll cell, pollen and seed) in L2, and the vascular bundle (cambium and pith) in L3, respectively [2]. A plant is considered to be a chimera if it has two or more genetic constitutions in its shoot apical meristem, and chimeras are classified into three types: sectorial, periclinal, and mericlinal [3,4]. Sectorial chimeras have a sector of all cell layers that is genetically different. Periclinal chimeras are chimeras in which one or more entire cell layer(s) is genetically distinct from another cell layer. Mericlinal chimeras have part of one or more layers that is genetically different.

In the genus *Citrus*, spontaneously arising chimeras have been reported previously; i.e., autogenous chimeras such as the "Suzuki wase" satsuma mandarin (*C. unshiu* Marcow.) and the "Thompson" grapefruit (*C. paradisi* Macfad.) [5,6] and graft chimeras such as the "Kobayashi mikan" [Natsudaidai (*C. natsudaidai* Hayata) + Satsuma mandarin], the "Kinkoji unshiu" [Kinkoji (*C. obovoidea* hort. ex Takahashi) + Satsuma mandarin], and the "Zaohong" navel orange ["Robertson" navel orange (*C. sinensis* Osbeck var. *Brasiliensis* Tanaka) + Satsuma mandarin] [7,8] are well known. All of these cultivars are periclinal chimeras. In recent years, production of synthetic graft chimeras has been successful [9,10], and many periclinal graft chimera cultivars have been registered in Japan. Furthermore, 2x+4x ploidy chimeras were produced by treating the apical meristems, undeveloped ovules, calluses, and protoplasts with antimitotic agents [11]. These were then used as breeding materials for triploid production in the genus *Citrus*.

In kumquats (*Fortunella* spp.), Yubeni which have a large fruit size and a high sugar content was discovered to be a bud mutation of the Meiwa kumquat. Yasuda et al. [12] demonstrated that Yubeni was a ploidy periclinal chimera with diploid cells in the outermost layer (L1) and tetraploid cells in the inner layers (L2 and L3) of the shoot apical meristem, and showed that the Yubeni increase in fruit quality was due to the tetraploidization in L2 and L3. Furthermore, this result showed that the 2x–4x–4x ploidy periclinal chimera kumquat could be used both for the parental line in triploid breeding and for direct domestication. However, since the ploidy chimera kumquat has rarely been reported [12], information on their characteristics is lacking. To prove the future usefulness of the ploidy chimera kumquat, it is necessary to evaluate more kinds and collect information.

Yahata et al. [13] produced many tetraploid Meiwa kumquats by applying a colchicine treatment to nucellar embryos. After they were grafted onto the trifoliate oranges [*Poncirus trifoliata* (L.) Raf.], they showed vigorous growth and they flowered and fruited for the first time, five years after budding. Nukaya et al. [14] examined whether these plants maintained tetraploidy, and found a 2x+4x ploidy chimera among their tetraploids. If this ploidy chimera is a ploidy peripheral chimera with the same histogenic layer as the Yubeni (layer constitution: L1–L2–L3 = 2x–4x–4x), it would be very useful in future kumquat breeding.

To clarify the ploidy level of each histogenic layer in the 2x+4x ploidy chimera, in the present study, we performed ploidy analysis by flow cytometry, cell observation using histological techniques and morphological characteristics in several tissues and organs of this ploidy chimera, and an evaluation of the reproductive organs of the chimera by crossing it with the diploid kumquat.

2. Materials and Methods

2.1. Plant Materials

The 2x+4x ploidy chimera, induced by applying a colchicine treatment to nucellar embryos of the Meiwa kumquat [14], was used in the present study. The original diploid Meiwa kumquat, the tetraploid induced by treating nucellar embryos of the Meiwa kumquat with colchicine, and the ploidy periclinal chimera mutant Yubeni, which originated from the Meiwa kumquat and has diploids in L1 and tetraploids in L2 and L3, were used as the control. These plant materials were grafted onto the trifoliate orange and grown in 45 L containers for approximately five years in the greenhouse of the Faculty of Agriculture, Shizuoka University, Shizuoka, Japan.

2.2. Confirmation of Ploidy Level by Flow Cyotometry

Approximately 50 mg segments of whole leaf, midrib, petal, filament, style, ovary, seed, juice sac, albedo, and flavedo were collected from each plant material. Their samples were chopped with a razor blade and blended for 5 min with a 2 mL buffer solution containing 1.0% (v/v) Triton X-100, 140 mM mercaptoethanol, 50 mM Na_2SO_3, and 50 mM Tris-HCl at pH 7.5, according to the preparation method of Yahata et al. [15]. An aliquot (550 μL) of each sample was filtered through Miracloth (Merck KGaA, Drarmstadt, Germany), and the filtrate was stained with 50 μL of 0.5 g L^{-1} propidium iodide (PI). The relative fluorescence intensity of the nuclear DNA was measured with a flow cytometry system (FCM, EPICS XL; Beckman Coulter, Inc., Pasadena, CA, USA) equipped with an argon laser (488 nm, 15 mW).

2.3. The Characteristics of Leaves, Flowers, Pollen, and Fruits in the Chimera

The characteristics of fully expanded leaves (e.g., blade size, thickness, guard cell size, guard cell density, and cell sizes) and flowers just before bloom (e.g., the size of the flower bud, petal, pistil, ovary, and pollen, and the number of petals and stamens) were measured. Guard cells and pollen grains were observed using a scanning electron microscope (Miniscope® TM3030Plus, Hitachi High-Technologies, Tokyo, Japan). The epidermises, palisade parenchymas, spongy parenchymas, vessels, and sieve tubes were used to measure the cell size. According to the methods of Nii et al. [16], their semi-ultra thin sections (1.5 μm) were cut using a glass knife before staining with methylene blue for histological examination under an optical microscope BX51 (Olympus Co., Ltd., Tokyo, Japan). Images were taken with a DP70 digital camera (Olympus). The weight, size, pericarp weight, the number of locules and seeds, soluble solids content (SSC), and titratable acidity (TA) of each kind of fruit were measured at the end of January. Each measurement used 20 samples.

Pollen fertility was evaluated by stainability and in vitro germination. Pollen stainability was estimated by staining the samples with 1% acetocarmine after squashing nearly mature anthers on a glass slide. In vitro germination of the pollen grains was performed on microscope slides covered with a 2 mm layer of 1% (w/v) agar medium containing 10% sucrose. Five stamens, each from different flowers, were rubbed on the agar medium, and the slides were then incubated for 10 h in a moistened chamber at 25 °C in the dark. Each test was evaluated from 500 grains with five repetitions.

2.4. Crossing for the Evaluation of the Reproductive Organs of the Chimera

The cross combinations are shown in Table 6. The flowers were pollinated immediately after emasculation and covered with paraffin paper bags. Seeds were collected from each mature fruit of all the crosses and were classified according to their size and shape into two groups: developed or undeveloped. After being numbered and weighed, both the developed and undeveloped seeds were cultured on Murashige and Skoog (MS) medium [17] containing 500 mg L^{-1} malt extract, 30 g L^{-1} sucrose, and 2 g L^{-1} gellan gum at 25 °C under continuous illumination (38 μmol m^{-2} s^{-1}). After germination, the seedlings were transplanted into vermiculite in pots and were transferred to a greenhouse. Ploidy analysis of the seedlings was performed by FCM using young leaves and chromosome observation using root tips according to the methods of Fukui [18] with some modifications.

3. Results

3.1. Confirmation of the Ploidy Level by FCM

The fluorescence intensities of all the examined tissues and organs of the diploid and the tetraploid Meiwa kumquats showed only diploid and tetraploid DNA values, respectively. In the 2x+4x ploidy chimera (Table 1, Figure 1), the whole leaf, petal, filament, style, ovary, and flavedo had both a diploid peak and a tetraploid peak, although the juice sac was diploid, and the midrib, seed, and albedo were tetraploid, respectively. Furthermore, the chimera showed the same fluorescence intensities as the Yubeni.

Table 1. Flow cytometric analysis of each organ and tissue in the diploid, the tetraploid, the Yubeni, and the chimera of the Meiwa kumquat.

	Leaf		Flower				Fruit			
	Whole	Midrib	Petal	Filament	Style	Ovary	Seed	Juice Sac	Albedo	Flavedo
Diploid	2x	2x	2x	2x	2x	2x	2x	2x	2x	2x
Tetraploid	4x	4x	4x	4x	4x	4x	4x	4x	4x	4x
Yubeni	2x+4x	4x	2x+4x	2x+4x	2x+4x	2x+4x	4x	2x	4x	2x+4x
Chimera	2x+4x	4x	2x+4x	2x+4x	2x+4x	2x+4x	4x	2x	4x	2x+4x

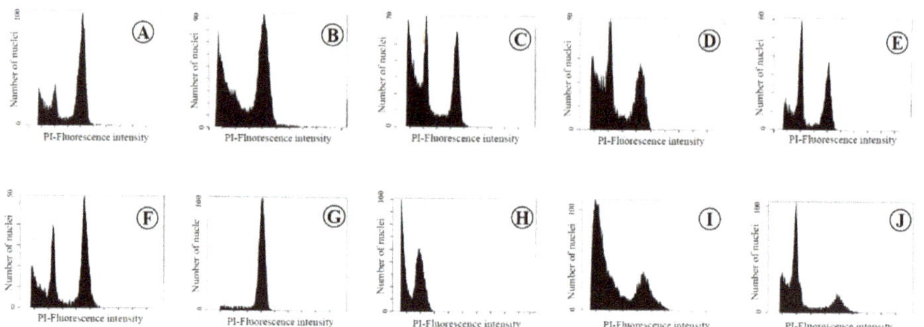

Figure 1. Flow cytometric analysis of each organ and tissue in leaves, flowers, seeds, and fruits of the chimera of the Meiwa kumquat. **A**: Leaf, **B**: Midrib, **C**: Petal, **D**: Filament, **E**: Style, **F**: Ovary, **G**: Seed, **H**: Juice sac, **I**: Albedo, **J**: Flavedo.

3.2. The Characteristics of Leaves, Flowers, Pollen, and Fruits in the Chimera

The morphological characteristics of the chimera were compared with those of the diploid, the tetraploid, and the Yubeni. The chimera had significantly rounder and thicker leaves as compared to those of the diploid, and it was mostly equal to those of the tetraploid and the Yubeni (Table 2, Figure 2A, Figure 3). On the other hand, the size and density of guard cells in the chimera was mostly equal to those of the diploid and the Yubeni (Figure 4). The epidermic cell size of the chimera showed the same value as the diploid and the Yubeni (Table 3, Figure 3). On the other hand, the palisade parenchyma and spongy parenchyma cells of the chimera were shown to be mostly the same sizes as the tetraploid and the Yubeni. Additionally, there were no differences in the cell sizes of the vessels and sieve tubes in midribs among the tetraploid, the Yubeni, and the chimera (Figure 5).

Table 2. Comparison of morphological characteristics of leaves in the diploid, the tetraploid, the Yubeni, and the chimera of the Meiwa kumquat.

	Ploidy Level	Leaf Blade (mm)		Shape Index of Leaf [y]	Leaf Thickness (μm)	Guard Cell (μm)		Guard Cell Density (No./mm^2)
		Length	Width			Length	Width	
Diploid	2x	75.2 a [z]	27.7 c	2.7 a	437.0 b	22.0 c	19.6 b	397.4 a
Tetraploid	4x	70.0 b	32.3 b	2.2 b	519.6 a	27.6 a	23.9 a	289.5 b
Yubeni	2x+4x	72 a b	32.9 b	2.2 b	532.3 a	24.2 b	19.3 b	419.3 a
Chimera	2x+4x	75.8 a	39.6 a	1.9 c	523.7 a	23.2 c	19.8 b	375.8 a

[y] Length of leaf blade/width of leaf blade; [z] different letters represent significant differences in Tukey's multiple test, 1% level.

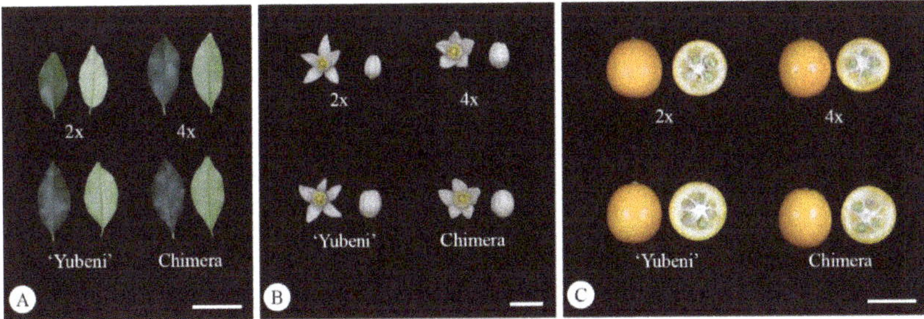

Figure 2. Comparison of the morphological characteristics of leaves (**A**, bar = 5 cm), flowers (**B**, bar = 1 cm), and fruits (**C**, bar = 3 cm) in the diploid, the tetraploid, the Yubeni, and the chimera of the Meiwa kumquat.

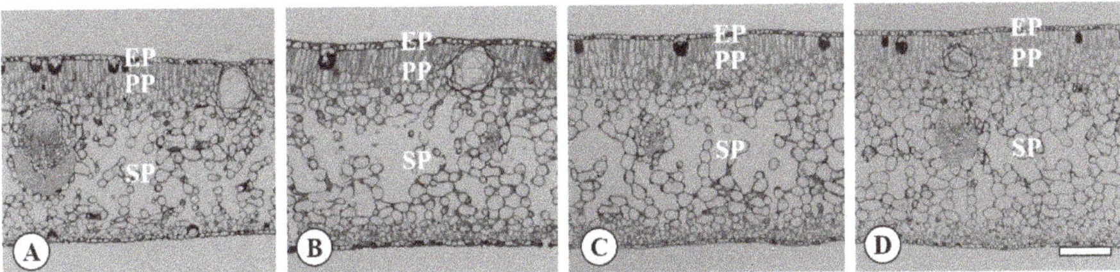

Figure 3. Transversal sections of leaves in the diploid (**A**), the tetraploid (**B**), the Yubeni (**C**), and the chimera (**D**) of the Meiwa kumquat. Bar = 100 μm. EP: Epidermis, PP: Palisade parenchyma, SP: Spongy parenchyma.

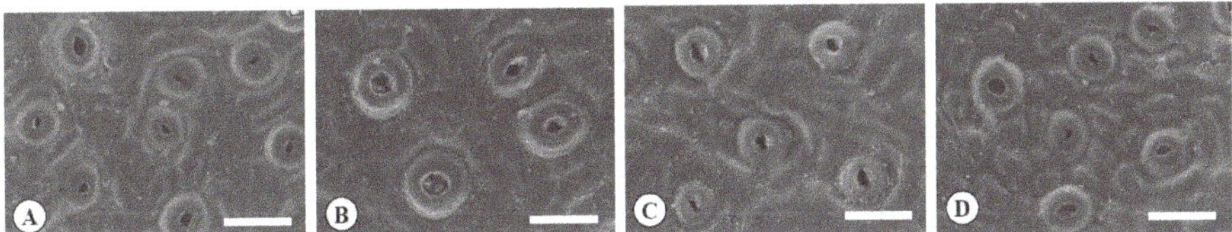

Figure 4. Scanning electron micrographs of guard cells in the diploid (**A**), the tetraploid (**B**), the Yubeni (**C**), and the chimera (**D**) of the Meiwa kumquat. Bars = 30 μm.

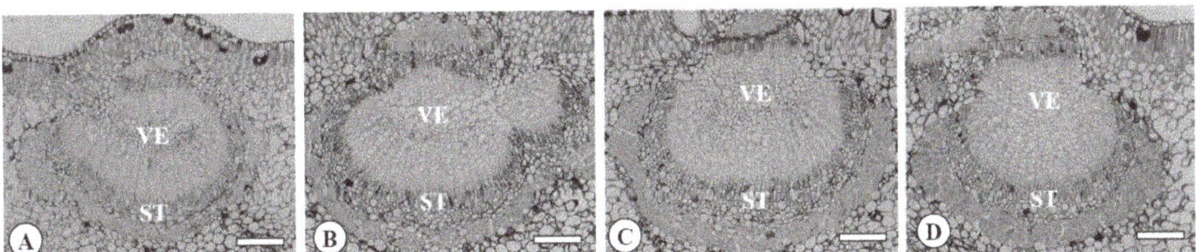

Figure 5. Transversal sections of midribs in the diploid (**A**), the tetraploid (**B**), the Yubeni (**C**), and the chimera (**D**) of the Meiwa kumquat. Bars = 100 μm. VE: Vessel, ST: Sieve tube.

Table 3. Comparison of the cell sizes of leaves in the diploid, the tetraploid, the Yubeni, and the chimera of the Meiwa kumquat.

	Ploidy Level	Epidermis (μm)		Palisade Parenchyma (μm)		Spongy Parenchyma (μm)		Vessel (μm)		Sieve Tube (μm)	
		Major Axis	Minor Axis	Major Axis	Minor Axis	Major Axis	Minor Axis	Major Axis	Minor Axis	Major Axis	Minor Axis
Diploid	2x	19.6 b [z]	13.5 b	27.5 b	8.9 b	35.5 b	25.5 b	18.6 b	16.9 b	19.0 b	15.1 b
Tetraploid	4x	27.4 a	15.9 a	41.7 a	13.0 a	60.4 a	40.2 a	26.2 a	22.6 a	22.0 a	18.0 a
Yubeni	2x+4x	16.0 c	11.9 c	42.6 a	11.5 a	54.2 a	34.9 a	26.3 a	22.0 a	24.4 a	18.8 a
Chimera	2x+4x	16.8 c	11.4 c	46.0 a	11.6 a	54.2 a	35.0 a	25.8 a	22.5 a	22.2 a	18.3 a

[z] Different letters represent significant differences in Tukey's multiple test, 1% level.

The chimera had significantly larger flower buds and ovaries as compared to those of the diploid (Table 4, Figure 2B). No difference in flower morphology was observed among the tetraploid, the Yubeni, and the chimera. The average size of the pollen grains from the chimera was larger than that of the grains from the diploid (Table 4, Figure 6). The pollen fertility of the chimera was significantly lower than that of the diploid and was about the same as those of the tetraploid and the Yubeni.

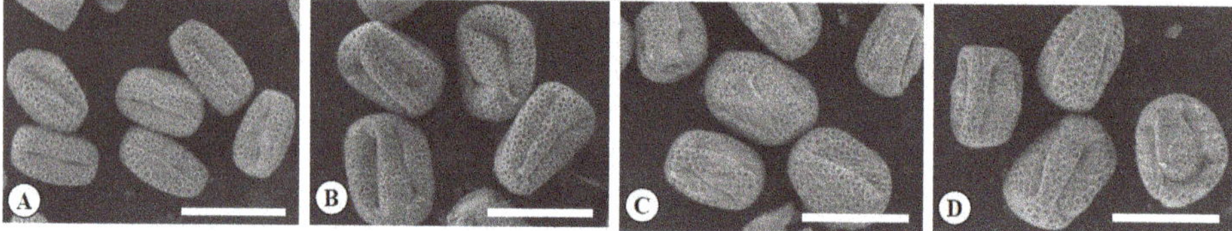

Figure 6. Scanning electron micrographs of pollen grains in the diploid (**A**), the tetraploid (**B**), the Yubeni (**C**), and the chimera (**D**) of the Meiwa kumquat. Bars = 30 μm.

There was no significant difference in the size of the fruits (Table 5, Figure 2C). However, the percentage of the pericarp weight per fruit in the tetraploid, the Yubeni, and the chimera was significantly higher in comparison to that of the diploid. The average number of seeds per fruit obtained from the diploid was 5.7, whereas that of the chimera was significantly lesser at 2.6. SSC in the pericarp and juice sac of the chimera was significantly higher than that of the diploid.

3.3. *Crossing for the Evaluation of the Reproductive Organs of the Chimera*

In order to evaluate the ploidy level in the reproductive organs of the chimera, crossing with the diploid Meiwa kumquat was carried out (Table 6). When the diploid was pollinated with pollen of the chimera, the frequency of developed seeds (23.1%) was lower than that of the self-pollinated fruit (86.8%). On the other hand, the frequency of developed seeds in the chimera was 86.7% when reverse-crossed, whereas that in the self-pollinated fruit was 89.2%. These developed seeds cultured on an MS medium germinated normally. The ploidy levels of these seedlings were confirmed by FCM analysis and chromosome observation (Table 6, Figure 7). Consequently, 8 and 12 triploid seedlings were obtained from crosses between the diploid and the chimera and from the reverse cross, respectively (Figure 8). Moreover, when the chimera was used as a seed parent, most of the seedlings were tetraploids.

Table 4. Comparison of morphological characteristics of flowers and pollen grains in the diploid, the tetraploid, the Yubeni, and the chimera of the Meiwa kumquat.

	Ploidy Level	Flower Bud (mm)		No. of Petal	Petal (mm)		Length of Pistil (mm)	Ovary (mm)		No. of Stamens	Pollen Grain (µm)		Shape Index of Pollen Grain [y]	Pollen Stability Rate (%)	Pollen Germination Rate (%)
		Length	Width		Length	Width		Diameter	Height		Length	Width			
Diploid	2x	8.8 c [z]	5.9 b	5.2	8.9 b	4.1 b	5.3 b	2.0 b	2.1 c	16.3	29.5 c	17.4 c	1.7 a	97.5 a	37.2 a
Tetraploid	4x	10.1 b	7.5 a	5.1	9.7 b	4.8 a	5.5 b	2.4 a	2.5 b	17.9	38.1 a	29.1 a	1.3 b	85.8 b	18.0 b
Yubeni	2x+4x	11.7 a	7.5 a	5.1	11.0 a	5.3 a	6.1 a	2.5 a	3.0 a	18.5	36.6 ab	25.2 b	1.4 b	72.4 c	12.5 b
Chimera	2x+4x	10.9 ab	7.1 a	4.9	10.8 a	5.1 a	5.6 b	2.4 a	2.5 b	16.9	35.2 b	25.4 b	1.4 b	80.2 bc	14.2 b

[z] Different letters represent significant differences in Tukey's multiple test, 1% level. [y] Length of pollen grain/Width of pollen grain.

Table 5. Comparison of morphological characteristics of fruits in the diploid, the tetraploid, the Yubeni, and the chimera of the Meiwa kumquat.

	Ploidy Level	Fruit Wt. (g)	Fruit (mm)		Shape Index of Fruit [z]	Pericarp Wt. (g)	Pericarp Wt./Fruit Wt. (%)	No. of Locules	No. of Developed Seeds/Fruit	No. of Undeveloped Seeds/Fruit	Developed Seeds/Fruit (%)	SSC (°Brix)		TA of Pericarp (%)
			Diameter	Height								Pericarp	Juice Sac	
Diploid	2x	22.2	33 6	35.9	93.7	12.6 b [y]	62.3 c	5.7	5.5 a	0.9	86.8 ab	16.3 c	14.0 b	0.31
Tetraploid	4x	22.7	33 7	35.1	96.1	15.9 a	74.5 a	6.4	3.8 b	1.5	74.2 b	19.8 ab	17.8 a	0.26
Yubeni	2x+4x	24.8	35 0	36.2	96.3	18.3 a	68.5 b	6.1	2.1 c	0	100 a	19.5 b	16.0 ab	0.30
Chimera	2x+4x	23.4	33 8	35.6	94.8	16.3 a	70.4 b	6.1	2.6 c	0.3	98.3 a	21.3 a	17.8 a	0.27

[z] (Diameter of fruit/height of fruit) × 100. [y] Different letters represent significant differences in Tukey's multiple test, 1% level. Wt.: weight.

Table 6. Seed content and ploidy level of the seedlings obtained from the reciprocal crosses between the diploid and the chimera of the Meiwa kumquat.

Cross Combination		No. of Fruits	No. of Seeds		No. of Developed Seeds/Fruit	Developed Seeds (%) [y]	Av. Developed Seed Weight	No. of Seedlings Examined	Ploidy Level		
Seed Parent	Pollen Parent		Developed	Undeveloped					2x	3x	4x
Diploid	Self-pollination	10	55	9	5.5	86.8	122.0	50	50	0	0
	Chimera	20	27	90	1.4	23.1	97.5	77	68	8	1
					*	*	*				
Chimera	Self-pollination	10	33	4	3.3	89.2	148.9	50	50	0	0
	Diploid	10	26	4	2.6	86.7	64.9	72	0	12	60
					*	NS [z]	*				

[z] NS: no significant difference. * mean is significantly different at 1% levels by *t*-test. [y] (No. of developed seeds/No. of seeds) × 100. Av.: average.

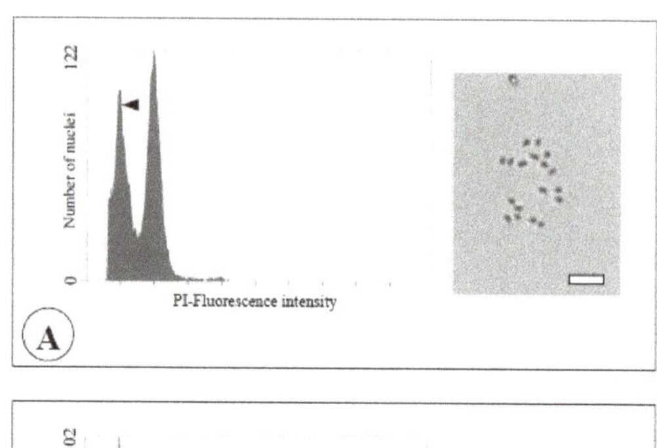

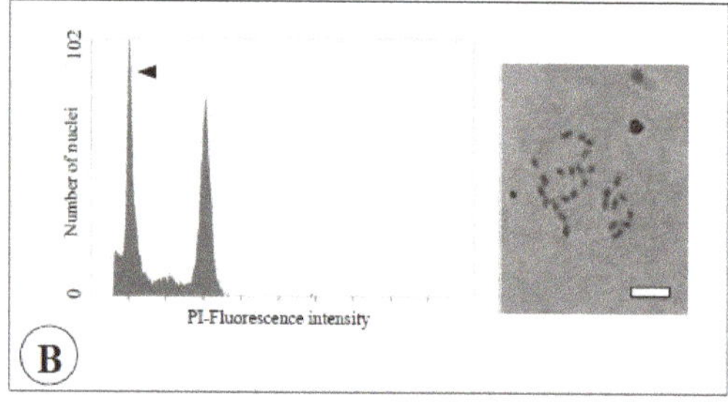

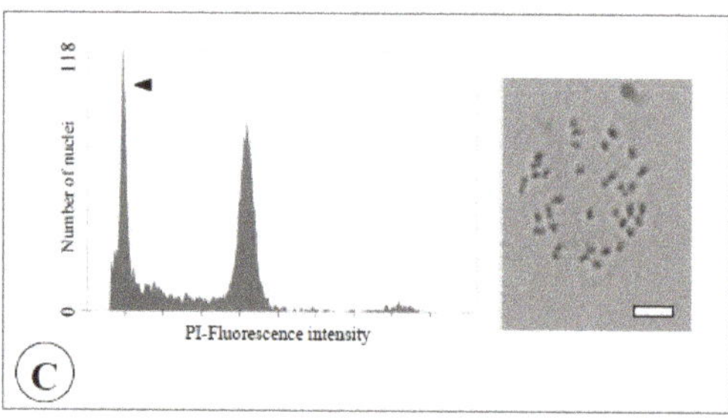

Figure 7. Flow cytometric analysis (left) and chromosome observation (right) of the seedlings obtained from reciprocal crosses between the diploid and the chimera of the Meiwa kumquat. Bars = 10 μm. **A**: Diploid (2n = 2x = 18). **B**: Triploid (2n = 3x = 27). **C**: Tetraploid (2n = 4x = 36). Arrows indicate the relative fluorescence of internal standard (the haploid pummelo (2n = x = 9, Yahata et al., [19])).

Figure 8. The triploid seedling obtained from the cross between the diploid and the chimera of the Meiwa kumquat (bar = 1 cm).

4. Discussion

In previous studies, the composition of each histogenic layer in periclinal chimeras such as the ploidy and the graft chimera was evaluated by morphological examination [11,12,20], histological observation [8–10], componential analysis [8], FCM analysis [12], and molecular biological techniques [7,9,10]. To clarify the ploidy level of each histogenic layer in the 2x+4x ploidy chimera, in the present study, flow cytometric analysis, cell observation, and morphological examination of several tissues and organs were carried out. Additionally, the ploidy chimera was crossed with the diploid kumquat to evaluate its reproductive organs of the ploidy chimera.

L1 is distinguished by the dermal system [3]. Especially, the size and the density of the guard cells were employed as the index to confirm the ploidy level in several plant species [19,20]. Epidermis cell size, guard cell size, and guard cell density of mature leaves in the chimera were mostly equal to those of the diploid control. This result showed that the chimera's ploidy level in L1 was diploid.

In the genus *Citrus* and its related genera, on the other hand, the juice sac is often used to analyze the origin of L1 [8,9,12]. Zhang et al. [8] demonstrated the origin of L1 by investigation of the carotenoid composition of the juice sac in periclinal graft chimera "Zaohang" navel orange composed of Robertson navel orange and Satsuma mandarin. Yasuda et al. [12] easily showed, by FCM analysis of the juice sac, that L1 ploidy level of the ploidy chimera Yubeni consists of diploid and tetraploid cells. In another study, however, Sugawara et al. [9] used RAPD analysis on a periclinal graft chimera composed of "Hamlin" sweet orange and Satsuma mandarin to show that cells derived from L1 and L2 were involved in the development of the juice sac. As reported by Nii and Coombe [21], the juice sac develops from epidermal cells and sub-dermal layers. In the present study, we could detect a single diploid peak just like a previous report by Yasuda et al. [12]. From here onwards, it has been considered that the juice sac in kumquats mainly differentiated from L1.

For L2 to differentiate into parenchymas and reproductive organs, mesophyll cells, pollen, and seeds are often used for the analysis of its origin [9,11,12]. The pollen of the chimera in the present study had as much size and fertility as the tetraploid and the Yubeni, and tetraploid and triploid progenies appeared in the cross between the chimera and diploid kumquat. Furthermore, the sizes of the subepidermal cells in the leaves of the chimera were significantly larger than those of the diploid, and were similar to those of the tetraploid and the Yubeni. For these reasons, it was considered that the L2 of the chimera was tetraploid.

Regarding L3, which is differentiated into cambium and pith, Yasuda et al. [12] presumed the ploidy level of L3 in the Yubeni by FCM analysis of the midrib. The FCM analysis of the midrib of the chimera showed tetraploidy. In the cell observations of the vessels and sieve tubes using histological techniques, furthermore, the cells of the chimera were the same size as the tetraploid and the Yubeni. Definitively, it was shown that the polyploidy of L3 in the chimera was tetraploid.

The chimera used in the present study was supposed to be a ploidy periclinal chimera, with diploids in the outermost layer (L1) and tetraploids in the inner layers (L2 and L3) of the shoot apical meristem. The morphological characteristics of the chimera were similar to that of the tetraploid Meiwa kumquat as previously reported [14,22,23]. Especially, it was reported that the fruit of these tetraploid kumquats had a thicker pericarp and a higher sugar content than the fruit of the diploid ones. The albedo, which is the main edible part of kumquats, differentiates from L2 [3], so Yasuda et al. [12] showed that the tetraploidization of L2 added to the superior fruit quality of the ploidy periclinal chimera Yubeni (diploids in L1 and tetraploids in L2 and L3). Because the chimera in the present study also had desirable fruit traits for kumquats, such as a thick pericarp, a high sugar content, and a small number of developed seeds, ploidy periclinal chimeras with tetraploids in L2 are very useful for kumquats. Furthermore, in the present study, triploid progenies were obtained from reciprocal crosses with the diploid kumquat. This result indicates that ploidy periclinal chimeras with tetraploids in L2 can be useful as parents for triploid breeding, where seedless fruits can be expected.

In conclusion, the 2x+4x ploidy chimera was confirmed as a ploidy periclinal chimera with diploids in L1 and tetraploids in L2 and L3. We plan to carry out the research not only on the utilization of the 2x+4x ploidy chimera for triploid breeding but also on the commercial growing of its chimera. Additionally, we need to develop an efficient production technique for the 2x–4x–4x ploidy chimera type.

Author Contributions: M.Y., A.T., H.M., K.Y., and H.K. conceived and supervised research. T.N., M.S., and M.Y. designed experiments. T.N. conducted most experiments. M.S. and T.O. assisted and performed some experiments. M.S. analyzed data and wrote the manuscript. M.Y., A.T., H.M., K.Y., and H.K. supervised the preparation of the manuscript.

Conflicts of Interest: The authors declare no conflict of interest.

References

1. Schmidts, A. Histologische studien an phanerogamen vegetationspunkten. *Bot. Arch.* **1924**, *8*, 345–404.
2. Frost, H.B.; Krug, C.A. Diploid-tetraploid periclinal chimeras as bud variants in citrus. *Genetics* **1942**, *27*, 619–634. [PubMed]
3. Cameron, J.W.; Frost, H.B. Genetics, breeding, and nucellar embryony. In *The Citrus Industry vol. II*; Reuther, W., Batchelor, L.D., Webber, H.J., Eds.; University of California Press: Berkeley, CA, USA, 1968; pp. 325–370.
4. Burge, G.K.; Morgan, E.R.; Seelye, J.F. Opportunities for synthetic plant chimeral breeding: Past and future. *Plant Cell Tiss. Organ Cult.* **2002**, *70*, 13–21. [CrossRef]
5. Cameron, J.W.; Soost, R.K.; Olson, F.O. Chimeral basis for color in pink and red grapefruit. *J. Her.* **1964**, *55*, 23–28. [CrossRef]
6. Nishiura, M.; Iwamasa, M. Chimeral basis for vegetative reversion in the Suzuki wase, an early maturing mutant of Satsuma mandarin. *Bull. Hortic. Res. Stn.* **1969**, *9*, 1–9.
7. Yamashita, K. Chimerism of Kobayashi-mikan (*Citrus natsudaidai×unshiu*) judged from isozyme patterns in organs and tissues. *J. Jpn. Soc. Hortic. Sci.* **1983**, *52*, 223–230. [CrossRef]
8. Zhang, M.; Deng, X.X.; Qin, C.; Chen, C.; Zhang, H.; Liu, Q.; Hu, Z.; Guo, L. Characterization of a new natural periclinal navel-satsuma chimera of *Citrus*: 'Zaohong' navel orange. *J. Am. Soc. Hortic. Sci.* **2007**, *132*, 374–380. [CrossRef]
9. Sugawara, K.; Wakizuka, T.; Oowada, A.; Moriguchi, T.; Omura, M. Histogenic identification by RAPD analysis of leaves and fruit of newly synthesized chimeric citrus. *J. Am. Soc. Hortic. Sci.* **2002**, *127*, 104–107. [CrossRef]
10. Zhou, J.; Hirata, Y.; Nou, I.S.; Shiotani, H.; Ito, T. Interactions between different genotypic tissues in citrus graft chimeras. *Euphytica* **2002**, *126*, 355–364. [CrossRef]
11. Yanagimoto, Y.; Kaneyoshi, J.; Furuta, T. Polyploid of pollen and fruit tissue, and utilization for triploid breeding in 2x−4x−4x ploidy chimera obtained by colchicine treatment of *Citrus*. *Hortic. Res.* **2017**, *16*, 249–257. [CrossRef]
12. Yasuda, K.; Kunitake, H.; Nakagawa, S.; Kurogi, H.; Yahata, M.; Hirata, R.; Yoshikura, Y.; Kawakami, I.; Sugimoto, Y. The confirmation of ploidy periclinal chimera and its morphological characteristics in Meiwa kumquat 'Yubeni'. *Hortic. Res.* **2008**, *7*, 165–171. [CrossRef]
13. Yahata, M.; Kashihara, Y.; Kurogi, H.; Kunitake, H.; Komatsu, H. Effect of colchicine and oryzalin treatment on tetraploid production of nucellar embryos in Meiwa kumquat (*Fortunella crassifolia* Swingle). *Hortic. Res.* **2004**, *3*, 11–16. [CrossRef]
14. Nukaya, T.; Ohta, T.; Yasuda, K.; Yahata, M.; Kunitake, H.; Komatsu, H.; Nii, N.; Mukai, H.; Harada, H.; Takagi, T. Characteristics in polyploid Meiwa kumquat (*Fortunella crassifolia* Swingle) induced by colchicine treatment to nucellar embryos and their utilization for triploid breeding. *Hortic. Res.* **2011**, *10*, 1–8. [CrossRef]
15. Yahata, M.; Nukaya, T.; Sudo, M.; Ohta, T.; Yasuda, K.; Inagaki, H.; Mukai, H.; Harada, H.; Takagi, T.; Komatsu, H.; et al. Morphological characteristics of a doubled haploid line from 'Banpeiyu' pummelo [*Citrus maxima* (Burm.) Merr.] and its reproductive function. *Hortic. J.* **2015**, *84*, 30–36. [CrossRef]
16. Nii, N.; Pang, C.X.; Ogawa, Y.; Cui, S.M. Anatomical features of the cell wall ingrowth in the cortical cells outside the endodermis and the development of the casparian strip in loquat trees. *J. Jpn. Soc. Hortic. Sci.* **2004**, *73*, 411–414. [CrossRef]
17. Murashige, T.; Skoog, F. A revised medium for rapid growth and bioassays with tobacco tissue cultures. *Physiol. Plant.* **1962**, *15*, 473–497. [CrossRef]

18. Fukui, K. Plant chromosome at mitosis. In *Plant Chromosome. Laboratory Methods*; Fukui, K., Nakayama, S., Eds.; CRC Press: Boca Raton, FL, USA, 1996; pp. 1–17.
19. Yahata, M.; Kunitake, H.; Yabuya, T.; Yamashita, K.; Kashihara, Y.; Komatsu, H. Production of a doubled haploid from a haploid pummelo using colchicine treatment of axillary shoot buds. *J. Am. Soc. Hortic. Sci.* **2005**, *130*, 899–903. [CrossRef]
20. Yanagimoto, Y.; Kaneyoshi, J.; Furuta, T.; Kitano, T. Characteristics of tetraploid and ploidy chimera induced by colchicine treatment in *Citrus*. *Hortic. Res.* **2012**, *11*, 449–457. [CrossRef]
21. Nii, N.; Coombe, B.G. Anatomical aspects of juice sacs of Satsuma mandarin in relation to translocation. *J. Jpn. Soc. Hortic. Sci.* **1998**, *56*, 375–381. [CrossRef]
22. Nukaya, T.; Sudo, M.; Yahata, M.; Nakajo, Y.; Ohta, T.; Yasuda, K.; Tominaga, A.; Mukai, H.; Kunitake, H. Characteristics in autotetraploid kumquats (*Fortunella* spp.) induced by colchicine treatment to nucellar embryos and their utilization for triploid breeding. *Sci. Hortic* **2019**, *245*, 210–217. [CrossRef]
23. Nukaya, T.; Yahata, M.; Suzuki, K.; Yasuda, K.; Kunitake, H.; Komatsu, H.; Mukai, H.; Harada, H.; Takagi, T. Fruit characteristics in autotetraploid Meiwa kumquat induced by colchicine treatment to nucellar embryos. *Hortic. Environ. Biotechnol.* **2009**, *50*, 188–190.

MDPI
St. Alban-Anlage 66
4052 Basel
Switzerland
Tel. +41 61 683 77 34
Fax +41 61 302 89 18
www.mdpi.com

Agronomy Editorial Office
E-mail: agronomy@mdpi.com
www.mdpi.com/journal/agronomy

www.ingramcontent.com/pod-product-compliance
Lightning Source LLC
LaVergne TN
LVHW070644170726
843515LV00004B/322

* 9 7 8 3 0 3 6 5 0 0 2 4 9 *